Studies in Computational Intelligence, Volume 81

Editor-in-chief
Prof. Janusz Kacprzyk
Systems Research Institute
Polish Academy of Sciences
ul. Newelska 6
01-447 Warsaw
Poland
E-mail: kacprzyk@ibspan.waw.pl

K. Sridharan
Panakala Rajesh Kumar

Robotic Exploration and Landmark Determination

Hardware-Efficient Algorithms and FPGA Implementations

With 50 Figures and 9 Tables

K. Sridharan

Panakala Rajesh Kumar

ISBN   978-3-642-09465-1   e-ISBN   978-3-540-75394-0

Studies in Computational Intelligence ISSN 1860-949X

Cover design: WMX Design GmbH, Heidelberg

Printed in acid-free paper

9 8 7 6 5 4 3 2 1

springer.com

Dedicated to our families

About the Authors

K. Sridharan received his Ph.D in the area of robotics from Rensselaer Polytechnic Institute, Troy, New York in 1995. He was an Assistant Professor at Indian Institute of Technology Guwahati from 1996 to 2001. He was a visiting staff member at Nanyang Technological University, Singapore in 2000-2001. Since 2001, he has been with Indian Institute of Technology Madras where he is presently an Associate Professor. He is currently visiting School of Computer Engineering, Nanyang Technological University, Singapore. He is a Senior Member of IEEE. His research interests include algorithms and architectures for robotics and image processing.

Panakala Rajesh Kumar received his Master's degree from Osmania University, Hyderabad, India. He was a Project Officer at Indian Institute of Technology Madras. He received his Ph.D from Indian Institute of Technology Madras in 2007. He is a visiting researcher at the Centre for High Performance Embedded Systems, Nanyang Technological University, Singapore. He is a Member of IEEE. His research interests include VLSI design and robotics.

Preface

Overview

Sensing and planning are at the core of robot motion. Traditionally, mobile robots have been used for performing various tasks with a general-purpose processor on-board. This book grew out of our research enquiry into alternate architectures for sensor-based robot motion. It describes our research starting early 2002 with the objectives of obtaining a time, space and energy-efficient solution for processing sensor data for various robotic tasks.

New algorithms and architectures have been developed for exploration and other aspects of robot motion. The research has also resulted in design and fabrication of an FPGA-based mobile robot equipped with ultrasonic sensors. Numerous experiments with the FPGA-based mobile robot have also been performed and they confirm the efficacy of the alternate architecture.

Organization and Features

Chapter 1 presents the motivation for the work described in this book. Chapter 2 surveys prior work on sensors and processors on mobile robots. Prior work on exploration and navigation algorithms is also described and the algorithms are examined from a VLSI point of view. Chapter 3 presents the details of design and fabrication of the FPGA-based mobile robot used for experiments reported in this book. Chapter 4 presents a hardware-efficient grid-based exploration strategy for dynamic planar environments. Chapter 5 studies the problem of constructing landmarks, in a hardware-efficient manner, using data gathered by the mobile robot during exploration. Chapter 6 presents a summary of the work described in this book and discusses extensions.

Programs have been developed in Verilog for exploration and other tasks. To give the reader a better feel for the approach, central modules in the Verilog code for exploration have been included in this book as an appendix. A list of suggested mini-projects is included as another appendix.

Audience

This book presents material that is appropriate for courses at the senior undergraduate level and graduate level in the areas of robotics and embedded systems. It is also suitable for researchers in the areas of VLSI-efficient algorithms and architectures. Practising engineers in the area of FPGA-based system design would also find the book to be immensely useful.

Basic familiarity with logic design and hardware description languages would be adequate to follow the material presented in this book.

Acknowledgments

The authors owe a word of thanks to many people who helped in various ways. The authors thank their families and friends for their support. Professor S. Srinivasan provided support and encouragement. Xilinx, Inc. and Digilent Inc. provided valuable support in the form of development boards for the experiments. Special thanks go to Dr. Thomas Ditzinger, Springer editor, for obtaining reviews for various chapters in this book. Thanks also to Ms. Heather King of Springer for assistance with the cover text. The authors thank the anonymous reviewers for their comments. The authors also acknowledge the support of Indian Institute of Technology Madras and Nanyang Technological University, Singapore.

K. Sridharan
Panakala Rajesh Kumar

Contents

Abstract

Robotic exploration of unknown environments has been of tremendous interest in the last decade. However, the literature has concentrated on static or semi-dynamic environments. Further, no restrictions on the size and energy consumption of the processing component have been imposed in the past.

This research presents new VLSI-efficient algorithms for robotic exploration, landmark determination and map construction in a *dynamic* environment where the geometry of the obstacles and their motion trajectories are not known *a priori*.

The input to the robotic exploration algorithm is a list of G nodes corresponding to locations determined using the robot's step size and the dimensions of the environment. P nodes accessible to the robot are identified and a map is obtained. The time complexity of the algorithm is $O(G)$. Special features of the algorithm include *parallel processing* of data from multiple ultrasonic sensors and the use of *associative memory* to efficiently keep track of the visited terrain nodes.

It is assumed that the exploration generates a subset P (of the G grid points) at which the robot can obtain sensor readings. Among P, there may be many points in close proximity and it is desirable to simply have representatives of sets of nearby points. We call them *landmarks*. An algorithm and architecture for construction of landmarks have been developed.

A mobile robot with eight ultrasonic sensors and an FPGA onboard has been fabricated (locally). The present design employs an XC2S200E (Xilinx Spartan series FPGA device). Experiments with the mobile robot are reported in this book.

1

Introduction

1.1 Motivation

Mobile robots and Automated Guided Vehicles (AGV) play a significant role in a variety of environments. These include domestic environments, industrial settings, military zones and others.

In general, mobile robots may operate with complete or incomplete information about the environment. With regard to the former scenario, a framework to solve what is known as the *Piano Movers' Problem* has been developed [1]. The framework assumes perfect knowledge of the environment. Construction of different types of geometric structures assuming specific models of the objects in the environment has also been studied in the domain of computational geometry by several researchers. Examples of geometric structures include visibility graphs [2] and Voronoi diagrams [3]. The geometric structures have been used in robotics for path planning.

The second kind, namely when complete information is not available, is of interest in a number of situations. Typically, it is difficult to know precisely the geometry of the objects for most robotic tasks. Various tasks for a robot without a priori knowledge of the environment have been studied. These include exploration and point-to-point navigation. With regard to exploration, considerable research has been done on learning the environment (as the robot moves) by means of various types of sensors mounted on the robot. Point-to-point navigation of a mobile robot in the presence of one or more obstacles when the geometry of the obstacles is not known a priori has also been extensively studied. It is assumed

K. Sridharan and P. Rajesh Kumar: *Robotic Exploration and Landmark Determination*, Studies in Computational Intelligence (SCI) **81**, 3–12 (2008)
www.springerlink.com

that the robot is equipped with sensors and it can use them for navigation.

One approach to navigation problems in unknown environments is based on exploration of the environment first and construction of a map. The map is generated in the free space [4]. For a given pair of start and goal positions for the robot, the task is to "retract" onto the map and move along the map as much as possible.

Another approach to navigation in an unknown environment involves taking the robot from the source to the destination without constructing a map. The robot may use proximity sensors as well as other sensors (for example, tactile sensors) for finding a path. An example of this is the BUG algorithm due to Lumelsky [5]. The algorithm in [5] is *sequential* and plans a path to a goal point from a start point efficiently. Enhancements to this have been developed [6].

Applications of autonomously operating robots are numerous. [7] documents a hospital transport robot to carry out tasks such as the delivery of off-schedule meal trays, lab and pharmacy supplies, and patient records. The robot in [7] is expected to navigate much like a human including handling uncertainty and unexpected obstacles. Ideas on sensor-based navigation developed in [5] are used in [7]. In recent years, a class of autonomous robots termed *humanoid robots* has been of tremendous interest. These robots are designed to perform a number of tasks (that humans traditionally do) such as providing assistance to the elderly and children. Further, a humanoid robot can also adapt to changes in the environment or itself and perform human-like movements, using legged locomotion.

The research described here broadly concerns hardware-efficient robotic exploration and mapping. Two scenarios are presented to motivate the discussion. The first scenario is illustrated in Figure 1.1. The environment consists of a number of static objects but the number, location and geometry of the objects are not known beforehand.

Robotic exploration of an environment such as the one shown in Figure 1.1 have been traditionally using an embedded computer (typically a laptop) on-board the robot. Embedded computers used on mobile robots for handling high-level tasks (such as

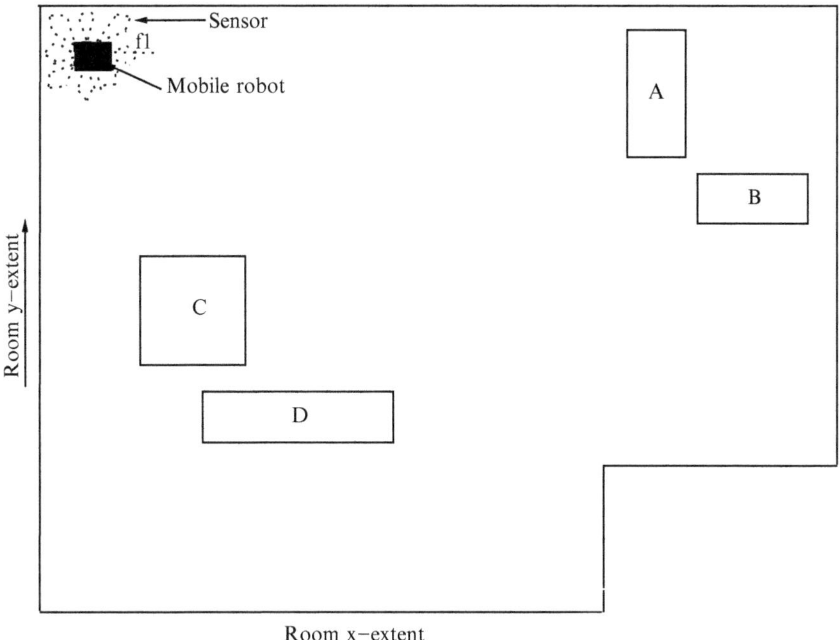

Fig. 1.1. Several static objects in an indoor environment; the number, geometry and location of the objects are not known

exploration, path planning, image processing etc.) consume power that is comparable to that of the motors [8]. Further, they involve a number of peripherals. The challenge here is to design alternative solutions that are area and energy-efficient.

The second scenario that this research attempts to tackle is shown in Figure 1.2. It involves machined parts (blocks/sheets). The challenge with respect to the scenario described by Figure 1.2 lies in the robot performing exploration and mapping without colliding with any dynamic or static objects and with "minimum" hardware.

In Figure 1.2, the block B moves between stations 2 and 3 slowly while the block A moves between stations 1 and 4 slowly. It is assumed that this happens in an automated way: as soon as block A gets cut (by the hacksaw type arrangement labelled CW) and moves towards station 4, another block is loaded from station 1 for a similar task. This process is assumed to take place continuously. Automated activity is also assumed to take place between stations 2 and 3. A robot must not attempt to cross the region

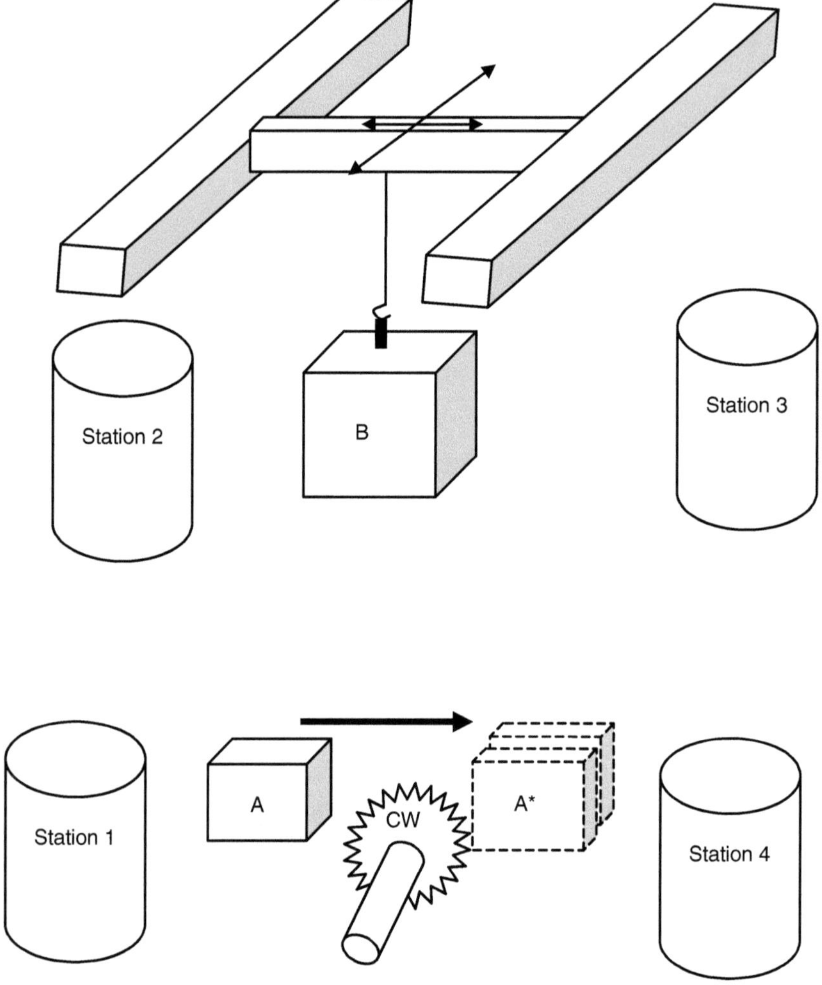

Fig. 1.2. *Simultaneous motion* involving parts machined/lifted

between stations 2 and 3 or the region between stations 1 and 4 during exploration/navigation. It is worth noting that an architecture based on one general purpose processor (as in a PC atop a mobile robot) can service multiple interrupts only as per some priority and further, it requires considerable (additional) support in the form of buffers and controllers.

1.2 Addressing the Challenges

The two scenarios presented call for an architectural alternative to general-purpose processors to meet the challenges. In particular, the first scenario calls for a processing style that facilitates use of a "small" battery on-board (the mobile robot). The second scenario suggests that an architecture with support for *parallel processing*: this would obviate the need for buffers and in general special peripheral support.

Towards achieving the objective of a time, space and energy-efficient solution, this work proposes a different model for processing sensor data. In particular, a model of computing based on Field Programmable Gate Arrays (FPGAs) is proposed. Development of hardware-efficient algorithms and implementing on a low-end FPGA on-board a mobile robot constitute the core tasks. To the best of our knowledge, design and implementation of hardware-efficient solutions has not been investigated previously for high level robotic tasks such as exploration and mapping.

Some comments about characteristics of FPGAs are in order. FPGAs are controlled at the bit level and when operating on narrow data items, FPGAs achieve much higher computational densities (measured in terms of bit operations per nanosecond) than general-purpose processors [9]. It is possible to implement parallel solutions with low space requirement. The ability to process data in parallel opens up a wide range of possibilities for handling many variations of problems dealing with dynamic environments. Further, FPGAs allow simple pin-level control, and a large number of sensors can be readily handled *in parallel* with user I/O pins.

An alternative to FPGAs with regard to implementation of parallel algorithms and designs is based on Application Specific Integrated Circuits (ASICs). ASICs can be developed to provide a high performance solution with low energy consumption. However, for the robotics application, it is advantageous to have a model where the user can design, test and modify algorithms. In this respect, FPGAs are preferred to ASICs and hence the former has been chosen for this research.

1.3 Architecture of an FPGA-based Robot

The architecture of a mobile robot with an FPGA incorporated into the system is shown in Figure 1.3. Several ultrasonic sensors are used to input data about distance to the "nearest" obstacles to the FPGA. The FPGA processes the information from the sensors *in parallel* and accordingly issues commands to the motor control unit of the robot. More details on the hardware and the mechanism are provided in the chapters to follow.

In addition to the ability to process data from multiple sensors in parallel, the architecture has other features. The architecture suggests a light-weight, area-efficient and low-power solution for robotic tasks: this is valuable for self-guided vehicles to complete exploration and mapping of a large area with less payload. Providing a low-energy/power solution for processing of data is important also for other reasons and in domains outside robotics: A number of motor vehicles today are equipped with navigational aids to avoid accidents and also enable the driver to know his current

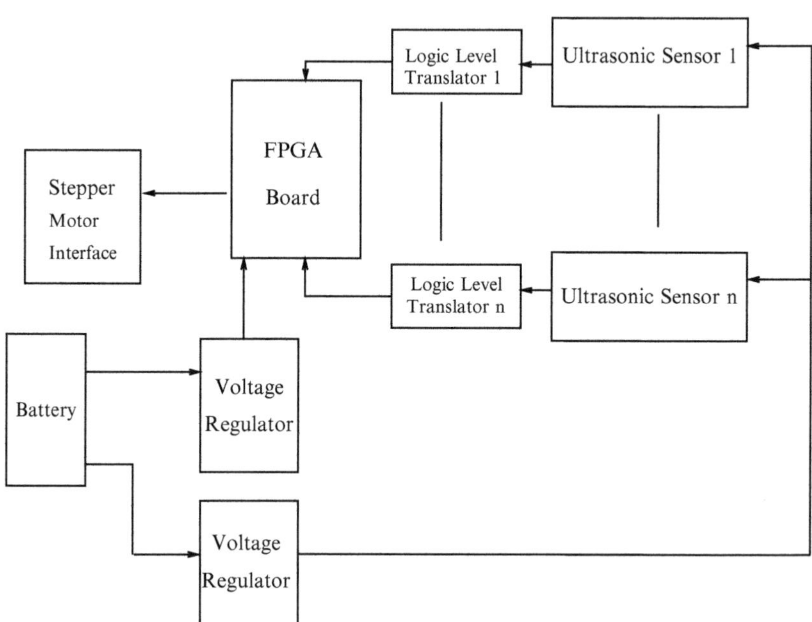

Fig. 1.3. Architecture of an FPGA-based mobile robot

location, the direction and distance to the destination. Most of these features draw power from a battery source so it is important to minimize energy consumption for handling sensor data.

Another aspect of interest with regard to exploration is storage of information about the environment. First, it is worth noting that there are many ways to explore the environment and depending on the pattern of exploration (that determines the number of locations at which the robot has to stop to gather sensor data), the amount of storage will vary. For systematic exploration of the environment, one can stipulate that the robot should be provided a list of locations that it should reach (or try to reach depending on the location of obstacles) and gather sensor data. This work investigates how best the data can be stored. Several approaches to store data are presented.

We now discuss other alternatives to general-purpose processors along with a brief commentary on their potential and applicability for robotic exploration. Various alternatives to general-purpose processors have emerged in the domain of computer architecture. Microcontrollers constitute one category of microprocessors optimized for embedded control applications. Such applications typically monitor and set numerous single-bit control signals but do not perform large amounts of data computations. Microcontrollers do not provide a means for implementation of *parallel algorithms*.

Digital Signal Processors (DSPs) constitute another alternative to general purpose processors. They are instruction set processors designed to perform operations such as addition, multiplication, shifting and adding etc. efficiently. The memory organization for DSPs is based on the Harvard architecture. However, they do not provide much support for implementation of parallel algorithms. Further, it is, in general, not possible to handle inputs in parallel as required in the robotic exploration application.

1.4 Contributions of this Research

The contributions of this work are the following:
(i) A testbed for FPGA-based processing of ultrasonic sensor data for exploration and mapping tasks has been developed. The entire platform consisting of the robot with sensors and other elements

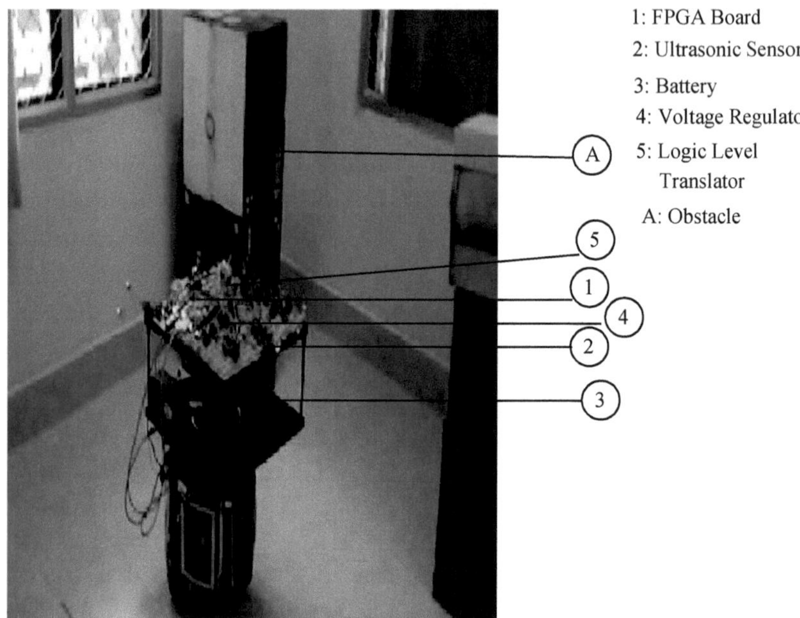

Fig. 1.4. Mobile Robot with eight ultrasonic sensors and FPGA

has been developed in-house. Figure 1.4 shows a picture of the robot with various accessories.

(ii) A new algorithm for robotic exploration in dynamic environments is proposed. The dynamic environments handled are comparable to those encountered in factories/workshops where humans/AGVs move between specific points at fairly low speed. The assumption with regard to moving objects is limited to a minimum value for the speed. No *a priori* knowledge of the geometries or the exact locations (of the objects) is assumed to be available. The input to the proposed algorithm is a list of G nodes corresponding to locations determined using the robot's step size and the dimensions of the environment. P nodes accessible to the robot are identified and a map is obtained. The time complexity of the proposed algorithm is $O(G)$. In particular, the strategy ensures repetitions of visits to nodes are minimized: this contributes to energy-efficiency. Also, special features of the algorithm include *parallel processing* of data from multiple ultrasonic sensors and the use of *associative memory* to efficiently keep track of the visited terrain nodes. This is accomplished along with an area-efficient design that accommodates the solution for a fairly large environment in a low-end FPGA device with a small number of system gates. Further, no extra hardware (such as a buffer) is required. Experiments with the robot confirm the time, space and energy-efficiency of the proposed approach.

(iii) The exploration process described may produce a large set of accessible points, particularly if the environment is not cluttered with obstacles. Not all of these points (locations) are truly required for most navigation tasks. In fact, by identifying an appropriate subset (of representative points), one can reduce the effort to find a feasible/shortest path from a point in the environment to another. In general, finding a representative subset is a computationally intensive task. The book presents a hardware-efficient algorithm for computing the representative subset (which we call as landmarks) of the set of accessible points. The algorithm has been mapped to hardware. The architecture comprises of components such as Content Addressable Memory and Wallace tree adders. Results of FPGA implementation show that the proposed design achieves high speed taking only a few milliseconds to identify approximately 50 landmarks (in a room of size roughly 4m x 4m).

It is worth noting that all of these are accomplished without using a personal computer (PC) as part of the loop. Sample Verilog programs for the exploration task are available from `http://www.ee.iitm.ac.in/~sridhara/Verilog_ExploreAlg`. Video files showing the implementation of the algorithm on our robot are available from `http://www.ee.iitm.ac.in/~sridhara/video_dyn_map`.

1.5 Organization of the Book

In the next chapter, we survey the literature on exploration and navigation by mobile robots.

In Chapter 3, we present the details of the FPGA-based robotic testbed developed locally for experiments reported in this book.

In Chapter 4, we present an algorithm and architecture for exploration of environments with static as well as dynamic objects.

In Chapter 5, we present a hardware-efficient algorithm for computing landmarks from accessible (sample) points identified during exploration (described in Chapter 4).

Chapter 6 summarizes the accomplishments and discusses extensions.

Appendix 1 presents the main Verilog modules for the exploration algorithm presented in Chapter 4.

Appendix 2 gives some suggestions for mini-projects based on the hardware designs and programs presented in the book.

2

Literature Survey

The survey is organized into sections covering prior work on various aspects presented in the subsequent chapters. Since a core component of the material presented in this book is description of an FPGA-based mobile robot developed for experiments, review of prior work on sensors and processors for mobile robots is first taken up. This is followed by a review of the literature on robotic exploration, landmark determination and navigation. Finally, we touch upon applications where FPGAs have been hitherto employed.

2.1 Sensors and Processors for Mobile Robots

Autonomous ground vehicles and mobile robots have been developed with different configurations and sensing capabilities since approximately the 1970s. One of the earliest wheeled robots was the SHAKEY [10]. The SHAKEY was based on a combination of on-board computers and a radio link to larger computers elsewhere. SHAKEY's primary sensor was a scanning camera. Around the early 1980's, the CMU Rover represented the state of the art in robotic mobility. The CMU Rover also had multiple general-purpose processors on-board and had provisions for connecting to a large remote computer.

While the robots described had sensing and planning capabilities, others which lacked elaborate planning capabilities were also simultaneously developed. Example are the Robart-I and Robart-II mobile robots [11]. Robart-I had a range of sensors, including ultrasonic sensors, infrared proximity detectors and tactile

K. Sridharan and P. Rajesh Kumar: *Robotic Exploration and Landmark Determination*, Studies in Computational Intelligence (SCI) **81**, 13–23 (2008)
www.springerlink.com   © Springer-Verlag Berlin Heidelberg 2008

feelers. Robart-II employed a microwave motion detector and a smoke detector besides infrared sensors. A good account of these robots is provided by [12].

Mobile robotic platforms with other types of processors are relatively recent. One example is the work reported in [13]. The authors in [13] report development of a mobile robot with two processors on-board: a TMS320F2812 DSP for implementing the control system for the robot and a general-purpose processor (as part of a laptop computer) for handling algorithms pertaining to navigation. Earlier work [14] has focussed on use of a digital signal processor for acquiring data from multiple ultrasonic sensors and processing. No experiment with an actual robot is reported in [14].

Many different sensors and instruments have been used on mobile robots. These include ultrasonic sensors, infrared sensors, vision sensors, tactile sensors, and encoders.

Ultrasonic sensors [15], [16] have been one of the simplest and widely used sensors for measuring distance to the nearest obstacle from a mobile robot. Design and development of circuits for distance measurement between a transmitter and receiver ultrasonic transducer have been extensively researched [17]. Approaches to correct various types of errors during distance measurement have also been developed [18]. Use of multiple receivers (along with one transmitter) has also been studied [19]. The exact point attaining minimum distance cannot be easily obtained using an ultrasonic sensor. This, however, is not a severe limitation for many applications in robotics.

Ultrasonic sensors have a range of approximately 3 metres and constitute the primary sensor on the mobile robot developed as a part of this book.

While sensors based on the ultrasonic, pulse-echo technique described in the previous paragraph have been the mainstay for collision avoidance systems in mobile robots, other alternatives that are not based on sound have also been examined by researchers [20]. In particular, where it is difficult to provide a medium for propagation of sound waves, light-based sensors have been explored. Some of these sensors are based on time-of-flight and triangulation techniques. In the mobile robot prototype developed in this research, three infrared sensors are used for detection of objects within a very small distance (less than 20 cm) from the robot and to cut off power (via a relay) to the motors driving the robot wheels. It is worth noting that light-based sensors are typically used to determine the presence of a "target" (or an object) rather than to measure distance to them.

Studies on tactile sensors and whiskers have also been made by researchers in the context of path planning for mobile robots. One example is the work described in [21]. Work on vision sensors for mobile robots also exists [22]. The authors in [22] discuss various aspects of mobile robot navigation using vision sensors. The work described in this research does not explicitly use tactile or vision sensors.

2.2 Robotic Operation in Known and Unknown Environments

A brief review of the literature on robotic operation in environments where the geometry of the objects is known beforehand (or a map is available) is presented. We then move on to study prior work on exploration and navigation in unknown environments.

2.2.1 Environment with Prior Knowledge of Object Geometries and Locations

It is worth noting that prior exploration (combined perhaps with other methods such as probing) would yield information on location and shape of objects (that function typically as obstacles for the robot during the course of navigation) in the environment. An important task then is to find a collision-free path for the robot from some initial position to a predefined destination. Several algorithms have been proposed for navigation in an environment when the geometry of the objects is completely known. These are based on notions such as the *visibility graph* [2] and Voronoi diagrams [1]. Much work has been done in the context of the *piano mover's problem* [1], [23]. The domain of computational geometry [3] has dealt with computational aspects of geometric problems dealing with polygons representing objects in the environment. While the solutions proposed for robotic navigation for different models of objects in the environment are interesting from a theoretical point of view, it is often difficult to obtain accurate *apriori* information on shapes and positions of the objects. The research reported in this book assumes that the geometry of the objects is not available.

2.2.2 Unknown Environments

Since the core of this book is on various tasks in unknown environments, we discuss work in different directions for this case.

Exploration and Mapping

One of the early efforts on *exploration* in unknown environments is [24]. The authors in [24] develop a method that is particularly appropriate for exploration and mapping in large-scale spatial environments. The authors examine the problem of errors in calculation of position for a mobile robot based on its encoders. The authors present a topology-based approach that relies on sensor signatures rather than on coordinates of a reference point on the moving robot with respect to a global coordinate frame. Several aspects pertaining to the difficulty in building a map using topological information are brought out. The authors present simulation results for map building.

Another piece of work on exploration is reported in [25]. The authors in [25] consider the notion of an agent that locates itself on a graph by matching nodes and the adjacency relationships between them. The approach assumes that the agent can label each node by depositing a marker at the nodes. The work reported is of an algorithmic nature. Since the focus of the work reported in [25] is not on implementation aspects, it is not clear how easily the ideas can be realized in hardware.

Exploration of environment for building specific types of geometric structures has also been performed [26], [27]. In [26], the authors present an incremental approach to construct a Voronoi diagram-like structure. In particular, the authors consider the problem of learned navigation of a circular robot through a two-dimensional terrain whose model is not *a priori* known. The authors present algorithmic details for (a) the visit problem where the robot is required to visit a sequence of destination points and

(b) the terrain model acquisition problem where the robot is required to acquire the complete model of the terrain. Experimental results are not available.

Construction of hierarchical generalized Voronoi graph for an unknown environment in two and higher dimensions is presented in [27], [28]. The authors in [27] present a scheme based on prediction and correction for incremental construction of the hierarchical generalized Voronoi graph. Planar and three-dimensional simulations are presented. Some experiments on a mobile robot equipped with a general-purpose processor and sonar are briefly described.

Construction of generalized local Voronoi diagram using laser range scanner data has been studied in [29]. The scheme developed in [29] is based on clusterization of scan points based on some property. In particular, clusterizations involving interdistance of successive scan points and distance to the nearest neighbor have been studied. It is worth noting that the above scheme is for a static environment and simulations are only presented.

Construction of other geometric structures based on sensor data has also been researched. In [30], the authors study the construction of a *Local Tangent Graph (LTG)* using range sensor data. The local tangent graph is then used for navigation. In particular, the local tangent graph helps to choose the locally optimal direction for a robot as it moves towards the destination point. Also, the robot uses the LTG for following the boundary of obstacles. The authors present simulation results for their scheme.

An approach to exploration based on visiting landmarks is presented in [31]. In [31], the work is based on a robot that maintains a list of list of all unvisited landmarks in the environment.

Considerable work has also been done on planning robot motion strategies for efficient model construction [32]. Work in this direction has been on finding a function that reflects intuitively how the robot should explore the space so that we have a compromise between possible elimination of unexplored space and distance travelled.

While most of the work in the area of robotic exploration has concentrated on static environments, there has been some work on semi-dynamic and dynamic environments. Mapping semi-dynamic environments where objects (such as chairs, tables) move periodically from one location to another has been studied in [33]. Some work on detecting changes in a dynamic environment for updating an existing map supported by computer simulations is described in [34]. Experiments with a mobile robot have been reported by some authors [35], [33].

More recently, an approach for planning exploration strategies for simultaneous localization and mapping in a *static environment* has been proposed [36]. The authors in [36] give a method to select the next robot position for exploration based on a novel utility function. The utility function defined in [36] combines geometric information with intensive usage of results obtained from perceptual algorithms. The outcome of the exploration is a multi-representational map made up of polygons, landmarks and a roadmap. Experiments with a real robot and simulations are presented in [36] but the focus is on using a general-purpose processor. Also, this work does not deal with *dynamic* environments.

Landmark Determination

The problem of selecting landmarks for path execution is addressed in [37]. In particular, the authors in [37] determine which landmarks a robot should detect and track at different parts of a given path so as to minimize the total cost of detecting and tracking landmarks. Experimental results on a mobile robot are not available. The authors present results of simulation based on implementations of a shortest path algorithm and a min-cost flow algorithm.

Localization of a free-navigating mobile robot called MACROBE is described in [38]. The authors in [38] use a 3D-laser range camera as the key sensor. Natural landmarks are extracted from the 3D laser range image and these are matched with landmarks predicted from an environmental model.

A method to learn dynamic environments for spatial navigation is proposed in [39]. The authors in [39] develop what are known as adaptive place networks and a relocalization technique. They then develop an integrated system known as ELDEN that combines a controller, an adaptive place network and the relocalization technique to provide a robust exploration and navigation method in a dynamic environment.

One of the common problems in landmark-based localization approaches is that some landmarks may not be visible or may be confused with other landmarks. The authors in [40] address this problem by using a set of good landmarks. In particular, the authors present an algorithm to learn a function that can select the appropriate subset of landmarks. The authors present empirical results to support their method. The authors in [41] develop a technique for on-line selection of stable visual landmarks under uncertainty of vision and motion.

A landmark-based approach for topological representation of indoor environments is proposed in [42]. Geometric features detected by exteroreceptive sensors of a robotic vehicle are grouped into landmarks. A complete representation of the environment is constructed incrementally as the robot relocalizes along its trajectory. Some robotic experiments using a HILARE-2bis and a laser range finder are also presented.

A technique based on training a neural network to learn landmarks that optimize the localization uncertainty is presented in [43]. Techniques to optimally select landmarks for mobile robot localization by matching terrain maps are presented in [44].

An approach to construct and tracking abstract landmark chunks from a video sequence is presented in [45]. Potential landmarks are acquired in point mode but aggregations of them are utilized to represent interesting objects that can then be maintained throughout the path. The approach has been tested on a video sequence of 1200 frames.

Navigation in Unknown Environments

We have so far studied literature on the *exploration* problem. We now briefly discuss literature on the *navigation* problem. With regard to *navigation* in unknown environments, one of the early approaches to sensor-based mapping (for navigation) is based on the notion of *occupancy grids* [15]. It is done by imposing a grid on the workspace. Each of the grid cells can have one of the following values: occupied, empty and unknown. The method proposed in [15] is restricted to a static unknown environment.

The occupancy grid technique has been applied to collision avoidance [46] and to learn shape models of objects in unknown environments [47]. In particular, [47] assumes a semi-dynamic environment where the objects are static over short time periods but move over longer periods (such as chairs moved in and out of a room). Some experiments with a *Pioneer* robot equipped with a general-purpose processor are described in [47]. Recently, variations to the occupancy grid approach have been proposed. One example is the *coverage map* defined in [48]. The coverage map is intended to overcome some of the difficulties encountered in the occupancy grid technique with regard to handling of walls and other obstacles.

Some work on navigation in busy dynamic environments also exists. The authors in [49] present the development of a robotic wheelchair and experiments in autonomous navigation in a railway station during busy periods. The authors employ a SICK two-dimensional laser range finder PLS200 and indicate that the real-time application calls for use of this expensive sensor. The authors employ a personal computer in their setup.

Hardware-directed Schemes

Since this book concerns digital hardware-friendly algorithms and FPGA implementations, we focus here only on reviewing prior digital hardware-based schemes for robotic planning and related problems. In general, not much is known on hardware-directed algorithms for robotic exploration and navigation.

A parallel algorithm and architecture for robot path planning are presented in [50]. The algorithm is based on wave propagations and the algorithm is mapped on to a systolic array architecture with N processing elements assuming the work area (for the robot) is divided into N x N discrete regions. The authors in [50] also indicate development of a prototype VLSI chip with five processors using 2-micron CMOS technology.

A parallel algorithm for collision-free path planning of a diamond-shaped robot among arbitrarily shaped obstacles, which are represented as a discrete image is presented in [51]. A cellular architecture and its implementation in VLSI using a 1.2 μm double-layer metal CMOS technology are also presented. No experimental results on an actual robot are available.

There exists some prior work on construction of certain geometric structures in hardware. The authors in [51] have proposed a cellular architecture for constructing a discrete Voronoi diagram on a binary image based on a d_4 metric (the d_4 distance between two pixels at coordinates (x_1, y_1) and (x_2, y_2) in an image is given by $\|x_1 - x_2\| + \|y_1 - y_2\|$). An extension of this work to construct the discrete Voronoi diagram based on the Euclidean distance metric in a binary image is presented in [52]. The approach in [52] is based on dilating each object iteratively and establishing connectivity between neighboring pixels belonging to the same dilated object at each iteration.

The authors in [53] consider the computation of generalized Voronoi diagrams on polygon rasterization hardware. The Voronoi diagram in [53] is obtained via a polygonal mesh approximation to the distance function of a Voronoi site over a planar rectangular grid of point samples. The design in [53] is focussed on a single processor solution and is supported by C++ simulations. More recently, the work in [53] has been extended in [54] for real-time path planning for virtual agents in dynamic environments. However, the focus of the work in [54] is on a PC-based solution with a special graphics card. The objective in [54] is to primarily obtain high speed and their focus is not on energy or space-efficiency.

It is also worth noting that in all of the above articles, there is no explicit sensor-based Voronoi construction. Other structures which have been explored from a hardware perspective include visibility graphs [55], [56], [57] and tangent graphs [58]. The authors in [56] present the notion of a virtual rectangle for path calculations. Work on definition of T-vectors in the context of mobile robot motion planning is presented in [55]. The approach presented in [55] is in the context of construction of reduced visibility graphs and is digital-hardware friendly. It is based on bit operations involving vertices of objects. However, the implementation of the approach in [55] is directed to a general-purpose processor.

In [57], the authors present a hardware-efficient scheme for construction of the complete visibility graph for (i) an environment with purely convex polygonal objects and (ii) an environment with non-convex polygonal objects. Results of FPGA-based implementation are also available.

In [58], the authors present a VLSI-efficient scheme for construction of the tangent graph. Results of synthesis using Synopsys Design Compiler 2001.08-SP1 with Compass Passport CB35OS142 standard cell library in 0.35 μm CMOS process are also presented.

In [55], [56], [57] and [58], the environment is assumed to be known.

2.3 FPGA-based Design

To the best of the our knowledge, there is no prior work on FPGAs for robotic exploration in dynamic environments. This section is primarily intended to give a feel for diversity of applications of FPGAs. The survey is not exhaustive. We have just attempted to cover some representative work.

FPGAs have been identified for use in space robotics [59] in view of the ability to get high performance solutions on a small device that may otherwise take up an entire board of parts.

A high-performance FPGA-based implementation of Kalman filter is reported in [60]. This is valuable in the context of probabilistic robotics with the focus on hardware-directed solutions. Codesign systems and coprocessors using FPGAs have been developed and employed in many applications in the last decade [61], [62], [63].

The authors in [64] study FPGA implementation of a fuzzy wall-following control scheme for a mobile robot. Work on handling other types of sensors on a mobile robot by an FPGA has also been reported recently [65]. The authors in [65] report an FPGA-based scheme for colour image classification for mobile robot navigation. No experiments on an actual report, however, are available in [65].

A proposal for a dynamic self-reconfigurable robotic navigation system is presented in [66]. The author in [66] presents an embedded system platform which is based on the partial reconfiguration capabilities of Xilinx Virtex-II Pro FPGA. The reconfiguration is suggested as a valuable feature in the context of image processing required on the mobile robot (in particular, to allow image loading and image matching to execute at the same time). The author presents some results pertaining to usage of the FPGA. Experimental results on navigation are not available.

An approach and architecture for dynamic reconfiguration on autonomous mini-robots is presented in [67]. The authors in [67] consider two application examples, one dealing with reconfigurable digital controllers and the second dealing with image processing to discuss their ideas. Two robotic platforms are also described.

An approach to incorporate partial reconfiguration in an environment with multiple mobile robots is presented in [68]. The authors in [68] present a case study involving autonomous fault handling and reconfiguration to show the behavior reconfiguration in a team of two robots (denoted by R_1 and R_2). In particular, R_1 and R_2 perform wall following initially. In the middle of this experiment, R_1's IR sensor fails so it notifies R_2. R_2, upon obtaining a resource list from R_1, formulates that Leader/Follower behaviour is appropriate and configures R_1 to follower behavior. The authors present snaps of actual point where reconfiguration takes place. They also present results of FPGA usage.

2.4 Summary

In this chapter, we have discussed the literature on sensors and processors used on mobile robots. The literature on exploration and navigation in unknown environments has then been discussed. Finally, applications of FPGAs have been mentioned.

3

Design and Development
of an FPGA-based Robot

3.1 Motivation

Currently, many robotic algorithms are implemented on general purpose processors. The emergence of Field Programmable Gate Arrays (FPGA) has given rise to a new platform for developing and implementing robotic algorithms for a variety of tasks. With FPGA devices, it is possible to tailor the design to fit the requirements of applications (for example, exploration and navigation functions for a robot). General-purpose computers can provide acceptable performance when tasks are not too complex. A single processor system cannot guarantee real-time response (particularly in the absence of considerable additional hardware), if the environment is dynamic or semi-dynamic. An FPGA-based robotic system can be designed to handle tasks *in parallel.*

An FPGA-based robot also improves upon the single general purpose processor/computer based robot in the following areas:

1. Enhanced I/O channels. One can directly map the logical design to the computing elements in FPGA devices.
2. Low power consumption compared to desktops/laptops.
3. Support for the logical design of the non-Von Neumann computational models.
4. Support for easy verification of the correctness of the logical design modules.

K. Sridharan and P. Rajesh Kumar: *Robotic Exploration and Landmark Determination*, Studies in Computational Intelligence (SCI) **81**, 25–34 (2008)
www.springerlink.com   © Springer-Verlag Berlin Heidelberg 2008

3.2 Overall Structure of the Mobile Robot

An FPGA based mobile robot platform has been fabricated (locally). The robot is designed to handle a payload of 10 kilograms. The present design employs the XC2S200E (Xilinx Spartan series FPGA device) which is adequate to implement the proposed exploration algorithm (and determination of landmarks for moderately-sized indoor environments).

The prototype provides features such as basic mobility on a plain surface and obstacle sensing. It consists of (1) two stepper motors to control the movement of two rear wheels (2) an array of eight ultrasonic sensors to measure the distance to the obstacles in vicinity of the mobile robot (3) one Digilent FPGA board (D2E-DI01) which consists of a Spartan XC2S200E device to process the ultrasonic sensor output values and issue commands to the stepper motor. The mobile robot has a corrugated aluminum base and moves on three wheels. The front wheel is a free rotating wheel and two rear wheels are independently driven by Sanyo Denki stepper motors. They move 1.8 degrees for each square pulse. The block diagram of the prototype is shown in Figure 3.1.

The different layers are shown in Figure 3.2. The bottom layer consists of electronics to control the movement of the rear wheels. The top layer consists of electronics to process the signals of the ultrasonic range sensors and to control the overall operation of the mobile robot.

Two versions of mobile robot prototype have been developed. In one version, bottom layer includes a microcontroller. In the other version, the top layer FPGA itself performs the tasks of the microcontroller as well.

The top layer consists of (1) Ultrasonic Range Finders (2) Logic Level Translators (3) Power Delivery System (4) FPGA Board. The bottom layer consists of Stepper Motor Interface and Stepper Motors.

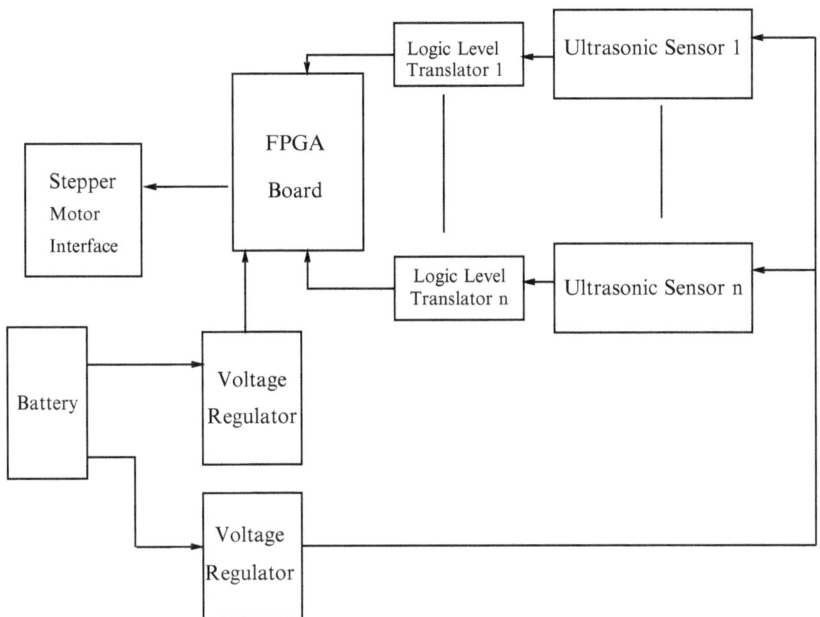

Fig. 3.1. Architecture of an FPGA-based mobile robot

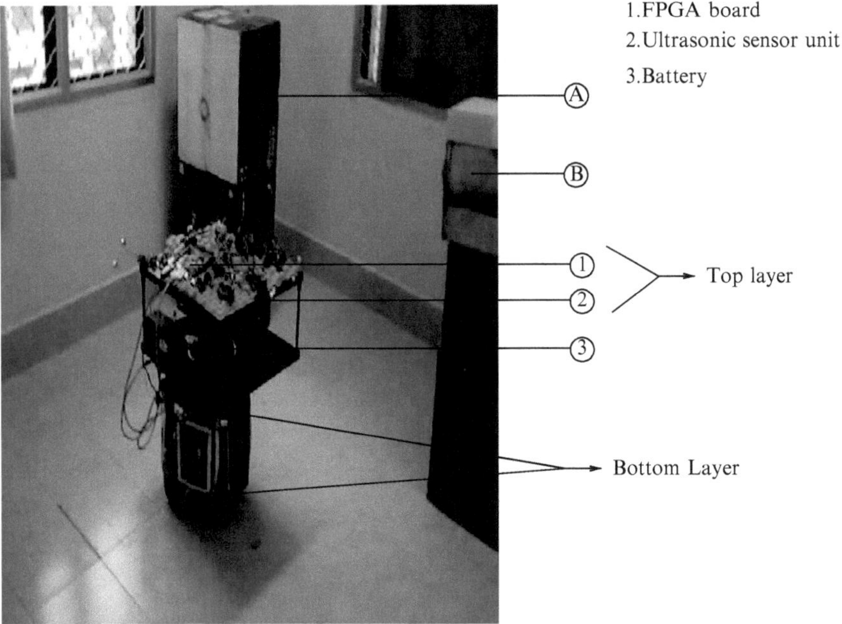

Fig. 3.2. The mobile robot

3.3 Design of Ultrasonic Range Finder

The ultrasonic range finder made in-house for our experiments
uses readily available components. The design is based on mod-
ifications to the one available from www.hobby-elec.org (website
last viewed on August 7, 2007). Typically, ultrasonic transmitter
and receiver pair devices work at around 40 kHz, as this is the res-
onant frequency of quartz crystals producing the tone. The range
finder consists of (1) an ultrasonic transmitter section and (2) an
ultrasonic receiver section as shown in Fig 3.3. The transmitter
section consists of two oscillators and an ultrasonic transmitter
driver. The first oscillator controls the on-off timings of the second
oscillator. The on-off timings are determined based on the range of
the ultrasonic range finder. With our ultrasonic sensor, distances

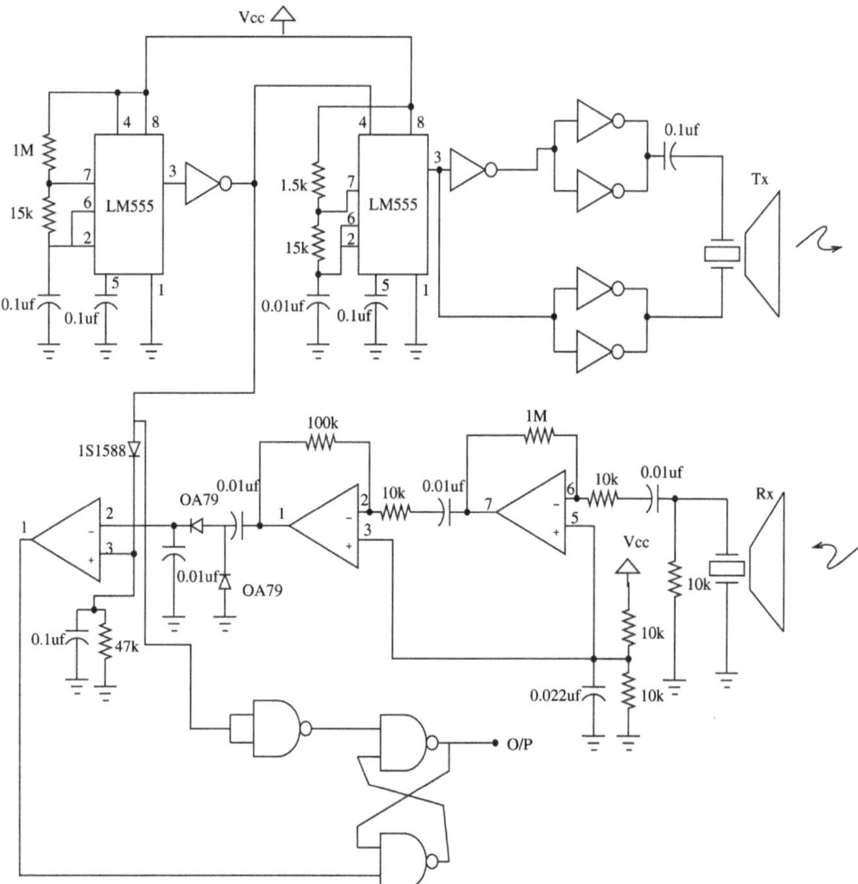

Fig. 3.3. Ultrasonic range finder

from 40cm up to approximately 2m can be reliably measured. The second oscillator generates a periodic square wave of 40 kHz. The driver consists of two inverters connected in parallel, so as to apply signals 180° out of phase on positive and negative terminals of the ultrasonic transducer, which in-turn doubles the strength of the signal at its input. The receiver section consists of an ultrasonic amplifier, a detector, a comparator and an SR flip-flop. The ultrasonic amplifier is a two stage amplifier and it amplifies the received ultrasonic echo signal by 1000 (60 dB).

The detector circuit is a half-wave rectifier which consists of two Schottky barrier diodes. The two inputs of the comparator are driven by the detector output and the first oscillator of the transmitter section respectively. The comparator output goes high only when the ultrasonic signal is detected. The S and R inputs of the SR flip-flop are driven by the first oscillator of transmitter section and the output of the comparator respectively. The output of the SR flip-flop goes high with the starting of transmission of ultrasonic signal and then goes low on the detection of the ultrasonic echo signal. The SR flip-flop's output pulse width is proportional to the time of flight of the ultrasonic signal.

3.4 Power Delivery to FPGA Board and Ultrasonic Range Finders

A 12V, 7AH sealed lead acid battery is the source of power for the ultrasonic sensor circuit and the FPGA board. To provide appropriate inputs for the ultrasonic sensor circuits and the FPGA board, voltage regulators are used. A separate voltage regulator is used for each type of hardware circuit board to provide isolation between logic and driver circuits. A 9V voltage regulator is used to provide constant voltage to ultrasonic range finder circuit board, while another 9V voltage regulator is used to power up the FPGA board. 9V and 5V voltage regulators are used to power up different parts of the stepper motor interface circuit board. The voltage regulator has been designed using LM317 IC.
The output voltage can be adjusted to the required voltage level by adjusting resistor R2 as shown in Fig 3.4.

3.5 Logic Level Translator

A logic level translator provides the level shifting necessary to allow proper data transfer in multi-voltage systems. The ultrasonic range finder gives out a 9V pulse but the FPGA chip can accept only a 3.3V pulse. A 74LS14 IC (which is a hex inverter with Schmitt trigger inputs) is therefore used as a level translator and is part of the interface. The 74LS14 IC connection diagram is shown in Fig 3.5.

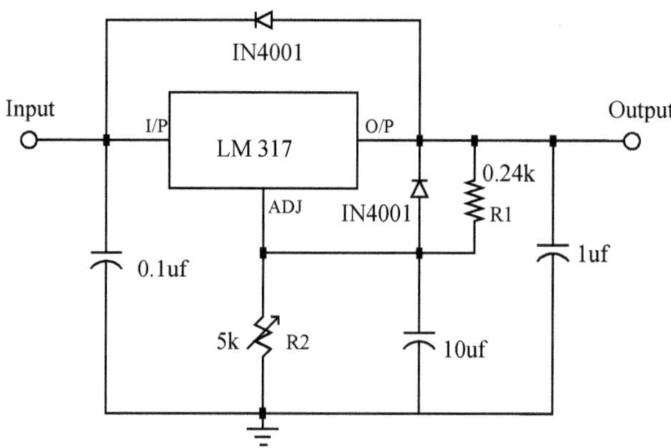

Fig. 3.4. Voltage regulator

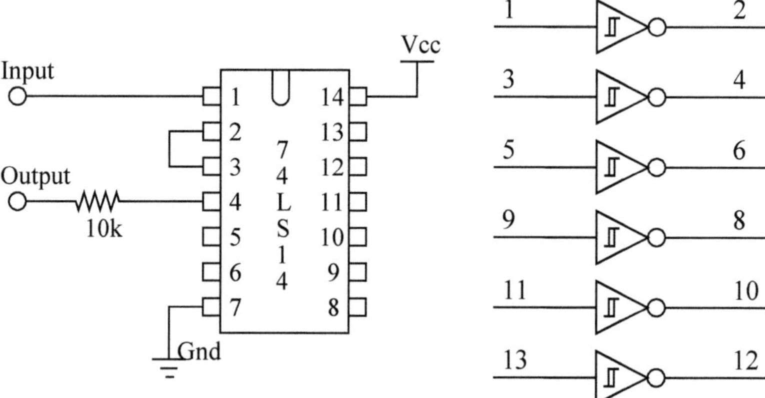

Fig. 3.5. Logic level translator

3.6 FPGA Board

The FPGA board consists of a Digilab 2E(D2E) board interfaced
with a Digital I/O (DI01) board. The D2E board consists of an
FPGA chip (Spartan-XC2S200E), an EPP capable parallel port
for JTAG based FPGA programming and a RS-232 port for serial
communication. The DI01 is an I/O board which can be attached
directly to a Digilab system board. It consists of push buttons,
switches (for input) and several LED displays (for output). Details
on D2E and DI01 boards are available at www.digilentinc.com.

3.6.1 Interface Modules

Interface modules for (1) Pulse Width to Distance Converter
(PWDC) to process ultrasonic range finder signal and (2) Universal Asynchronous Transmitter(UAT) have been developed in
Verilog and mapped on to the FPGA to issue commands to the
stepper motor driver. The internal details of the FPGA interface
are shown in Fig 3.6.

3.6.2 Pulse Width to Distance Converter (PWDC)

The PWDC consists of a 16-bit counter and a clock divider. The
16-bit counter has been chosen to maintain a high level of precision

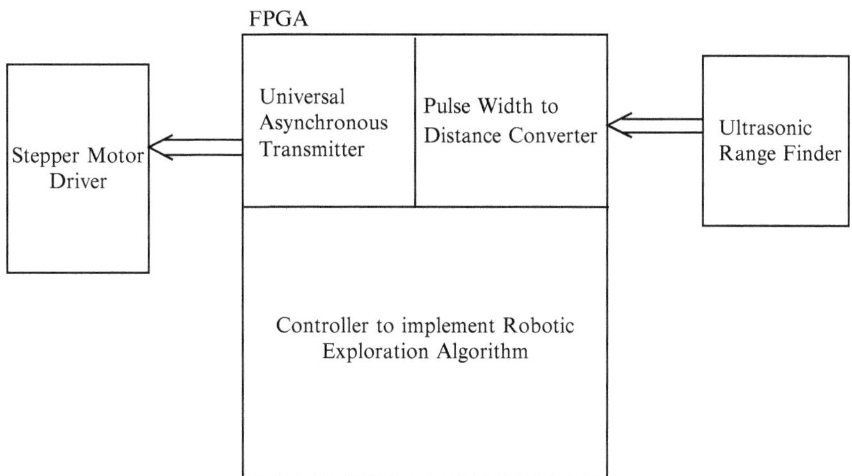

Fig. 3.6. Interfacing FPGA and other units

in conversion, i.e the distance value is obtained in millimetres. The counter is driven by the output signal of the clock divider, whose frequency is decided based on speed of the ultrasonic signal and precision of conversion.

3.6.3 Universal Asynchronous Transmitter (UAT)

The UAT consists of a controller, a clock divider and a parallel to serial converter. The controller receives command words which have to be transmitted serially to the stepper motor driver through RS-232 serial communication. The output signal frequency of clock divider is set at 10 kHz to transmit data at the rate of 10 Kbps. The parallel to serial converter begins serial transmission upon receiving a START signal from controller and then sends back an ACK signal upon completing the transmission of data.

3.7 Description of Stepper Motor Interface

The motors are driven by batteries which are on the robot. No separate tethered power source is required. The stepper motor interface consists of an L297 [69], which is a stepper motor controller and an L298, which is a bridge driver. This is shown in Fig 3.7.

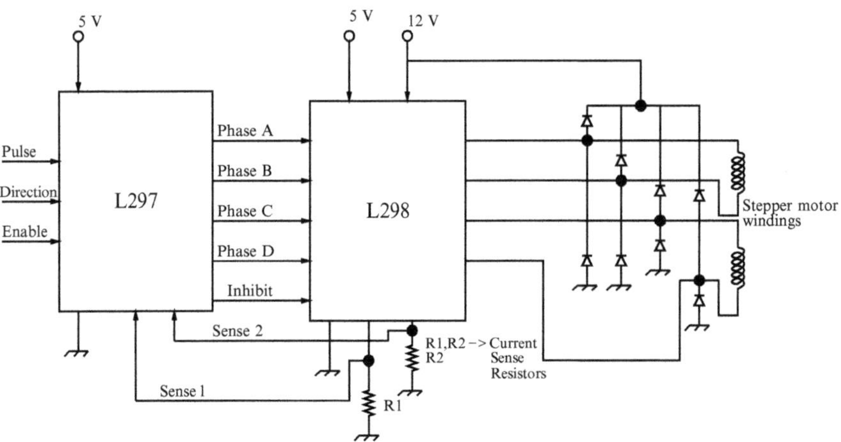

Fig. 3.7. The stepper motor interface

The L297 consists of a translator and an output logic circuit. The translator generates the required phase sequences upon receiving a set of control signals from the system controller, i.e either from a FPGA chip or from a microcontroller. These phase sequence signals are further processed by the output logic circuit which consists of a chopper and an inhibitor. The chopper controls the load currents to obtain good speed and torque characteristics. The inhibitor speeds up the current decay when a winding is switched off. The L298 is a monolithic bridge driver. A full bridge configuration to drive stepper motor in both directions is shown in Fig 3.8.

In Fig 3.8, Vinh denotes the inhibitor voltage while C and D denote the phase C and phase D voltages. All these are applied by the L297. The Table 3.1 depicts how the motor is being run.

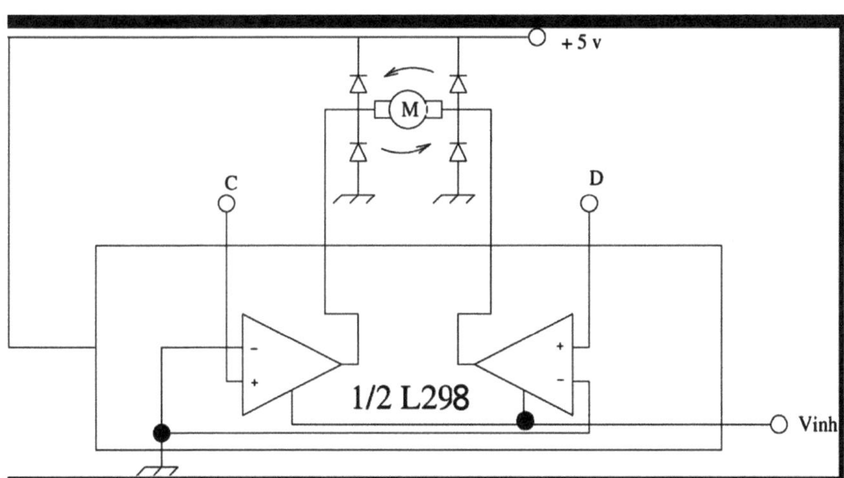

Fig. 3.8. Bidirectional driver of stepper motor

Table 3.1. Stepper Motor Control Signals (L = Low, H = High, X = Don't Care)

Inputs		Motor Action
Vinh = H	C = H, D = L	Turn Right
	C = L, D = H	Turn Left
	C = D	Fast Motor Stop
Vinh = L	C = X, D = X	Free Running Motor Stop

3.8 Summary

In this chapter, the details of the mobile robot fabricated in our laboratory have been given. The Xilinx device XC2S200E has been successfully interfaced and tested with ultrasonic range finders and stepper motor drivers.

Hardware-Efficient Robotic Exploration

4.1 Introduction

As pointed out in Chapter 1, the focus with regard to robotic exploration has been on use of a *general-purpose processor* for handling sensor data and performing the required computations. As a consequence, additional hardware (in the form of buffers) has become necessary. This chapter examines the use of FPGAs, an architecturally-efficient alternative to general-purpose processors, for the exploration task.

In this chapter, we present a new VLSI-efficient algorithm for robotic exploration and mapping in a *dynamic* environment where the geometry of the objects or their motion trajectories are not known *a priori*. The challenge here lies in the robot performing exploration using compact on-board electronics and avoiding collision with any dynamic or static obstacle. The simultaneous motion of dynamic objects makes the exploration task difficult.

Our approach to solve the exploration and mapping problem is based on assumption of a grid of G nodes in the exploration space where G is determined merely using the dimensions of the room and the robot's step size. It is worth noting that not all of the G nodes may be accessible: some may be occupied by a static or dynamic obstacle. In order to enable the robot to complete systematic exploration, it is further assumed that there is a connected set of free nodes (not occupied by a static or dynamic obstacle) and the robot can *identify* them based on sensor readings and move from one such node to the next. P nodes accessible to the robot are identified: any non-accessible node NA may be either occupied

K. Sridharan and P. Rajesh Kumar: *Robotic Exploration and Landmark Determination*, Studies in Computational Intelligence (SCI) **81**, 35–62 (2008)
`www.springerlink.com`   © Springer-Verlag Berlin Heidelberg 2008

by static/dynamic obstacles or some static obstacle occupying for example a diagonal node may prevent NA from being reached. A map that does not cut across any of the non-accessible nodes is then obtained.

The proposed algorithm has *linear* time complexity. Along with the optimal time complexity, the algorithm has other features. Special features include *parallel processing* of data from multiple (low-cost) ultrasonic sensors and determination of neighbors accessible to a specific grid point in the terrain *in parallel.*

An efficient FPGA implementation of the design tailored to use Block Random Access Memories (BRAMs) on a Xilinx XC2S200E is also presented. The utilization of Block RAMs for handling memory blocks in the design helps to reduce the reliance on slice flip-flops thereby a large number of sensors can be handled. Further, utilizing BRAMs for large storage helps to get energy savings [70].

In the next section, we present the assumptions and terminology required for the development of the algorithm. Subsequent sections describe the algorithm, architecture and implementation aspects.

4.2 Assumptions and Terminology

The algorithm is designed for a robot with eight ultrasonic sensors (pointing along $E(east)$, $NE(northeast)$, $N(north)$, $W(west)$, $NW(northwest)$, $S(south)$, $SW(southwest)$, and $SE(southeast)$). Motions of the robot are restricted to be along four directions, namely E, W, N and S. Eight sensors have been chosen keeping in view the four directions of motion for the robot as well as the possibility of obstacles located diagonally (for eg. NW of the present robot location) obstructing movement of the robot. Sensors pointing along NE, NW, SE and SW are used to gather information about obstacles located 'diagonally'. These eight sensors are well-separated to avoid interference of signals between adjacent ones. In order to ensure uniform exploration of the environment, equally-spaced grid points (nodes) have been chosen. The robot, based on sensor information about obstacles, will determine if it can move to a node that is one step away from its present position.

The robot starts from a corner designated as the origin and moves along E-direction. As illustrated for factory environments, motions of dynamic objects are typically east/west or north/south. Further, they move slowly (nominal speed of 0.5 m/s is assumed). It is worth noting that this enables setting a maximum waiting time (in experiments, this is in the order of a few seconds). Any other motion should not violate the requirement that there is a connected set of free nodes (not occupied by any static/dynamic object) to enable the robot to complete systematic exploration. In place of humans, parts to be machined could move between stations. The dynamic activity is assumed to take place continuously so the robot during its exploration can capture this regardless of the time at which the exploration begins.

The starting point is also assumed to be not occupied by any obstacle. Only initial/starting point's coordinates are provided to the FPGA, not that of all the grid points. While the exact point hit by a ray from an ultrasonic sensor is difficult to determine, it is worth noting that for the exploration and mapping considered here, this information is not explicitly required. Obstacles are assumed to be not *anechoic* and further, obstacles do not have any openings that may prevent reception of signal by the transducer setup. It is also assumed that obstacles are isolated enough: otherwise, energy may return from multiple surface reflections resulting in erroneous distance values. Dead reckoning is used throughout to obtain information on the position of the robot on its course.

4.3 The Proposed Algorithm

4.3.1 Key Ideas

The algorithm may be perceived as a priority search. The priority for visit of nodes adjacent to a given node (grid point) is assigned as follows: E, W, N, S. The rationale for this assignment is as follows: The robot is positioned to move along E to begin with and it will continue to move east unless an obstacle forces a change of direction. When the robot cannot move (can no longer move) along east (i.e., there is an obstacle either along E or at the diagonally located node), the next direction chosen is west and so on. The starting point's coordinates are provided to the processing unit (FPGA in this case). Based on dimensions of the environment and the step size of the robot, the maximum number G of grid points is obtained. In order to detect dynamic obstacles crossing specific grid points, the robot waits at each node for D_m/s_{min} time where the maximum distance between two opposite "corners" of the environment is denoted by D_m and the minimum speed of obstacles is denoted by s_{min}. During D_m/s_{min}, the sensor inputs are taken every t milliseconds where t depends on the speed of sound and the distance to the next node (with respect to this algorithm). In our case, t is 5 milliseconds (since we monitor activity near the adjacent node that is 40 cm away). Each of the eight sensors monitor a neighbor grid point. Depending on the obstacles detected, the robot decides its course.

To successfully handle multiple moving objects (i.e., to "catch" the simultaneous motion), we perform processing of data from sensors *in parallel*. It is worth noting that there is **no** initial storage of possible grid point coordinates for all the G points. As a node is visited, its index is obtained from one of four *adjacency memory blocks*, denoted by APX, AMX, APY and AMY, depending on the direction (east, west, north, south) in which the adjacent node is located. The storage of indices is one-dimensional. The index is pushed on to a stack and inserted into a Content Addressable Memory (denoted by CAM). Index computation uses information on the number of rows: if the environment has 128 grid points (with 16 points per row), then the index of the node along north visited from node whose index is 1 would be 17. The size of the stack is determined from G. The stack facilitates bringing back the mobile robot to the previous node if at the present node, all the neighbours have been marked as visited. The CAM helps to check if a node has been visited or not. Since a dynamic environment consists of nodes that are temporarily occupied and these are also *not* part of the final map, the CAM also stores information on these temporarily occupied nodes.

Coordinates of a node visited are recorded in two separate memory blocks, called as *Visited grid point_x memory* and *Visited grid point_y memory*. Sensor values (distances) of the corresponding nodes are stored in four memory blocks denoted by DPX, DMX, DPY and DMY (where DPX and DPY stand for distance to nearest obstacle for sensors pointing along X and Y

directions respectively; similar meaning is associated with DMX and DMY). The coordinates of visited nodes are obtained using previously visited node's coordinates, robot's step size and direction of motion of the robot. The complete architecture is presented in section 4.4.

4.3.2 Pseudo-Code for the Proposed Algorithm

The robot's reference point is aligned with the starting location. The environment is assumed to have G grid points (and $P \geq 2$; typically much more than 2). The step size for the robot is denoted by d. Storage of visited point coordinates is in memory blocks denoted by *Visited grid point_x memory* and *Visited grid point_y memory*. It may be noted that these memory blocks contain information on the P accessible grid points. Four registers denoted by $N_i, i = 0, \cdots, 3$ are used to store indices of E, W, N and S neighbors for a given node : when a node does not have these four neighbors (boundary is encountered), the corresponding register stores a value denoted by FF. FF is assumed to be larger than the maximum possible one. The registers get the indices from adjacency memory blocks. A four-element vector denoted by *visit_tempoccupied* is used to temporarily assign values of 1 or 0 depending on the status of a neighbor (along E, N, W and S) for the current node (node at which the robot is). In the algorithm, we use the term *temporarily occupied grid point* to refer to a node that is occupied by humans or chairs/tables etc. (that are likely to be shifted later).

Algorithm *Explore_Dynamic_Environment*:

Step 1: Initialize (i) robot location (ii) distance counter values of ultrasonic sensor circuit interface (iii) grid point's index and (iv) *Visited grid point_x memory* and *Visited grid point_y memory* with coordinates of starting location for the robot. Reset a match-found flag associated with CAM.
Step 2: Push grid point's index on to the stack. Also, insert grid point's index into the CAM. Obtain the inputs from the eight sensors *in parallel* to determine the neighbors (E, N, W, S) that can be visited. Repeat sampling of sensory inputs every t milliseconds

(actual value depends on the speed of sound and the distance to the next node) for a total time of D_m/s_{min}.

Step 3: Insert into the CAM the indices of temporarily occupied neighboring grid points (if there are any).

Step 4: Store the distance values (for the sensors pointing along E, W, N and S) in memory blocks DPX, DMX, DPY and DMY. Get the neighbor grid point indices *in parallel* from adjacency memory blocks APX, AMX, APY and AMY (and store in $N_i, i = 0, \cdots, 3$).

/* check neighbors as per priority */

Step 5: for i = 0 to 3 do

Input N_i to CAM

if match-found flag (of CAM) is set

visit_tempoccupied[i] = 1

else

{

visit_tempoccupied[i] = 0

if((visit_tempoccupied[i] == 0) & ($N_i \neq FF$)

go to step 7

}

end for

Step 6: Pop previously visited grid point's index g_{pv} (that is, use parent information to go back) and transfer the robot to g_{pv}. If the stack is empty, go to Step 8. Else, test if g_{pv} has any unvisited neighbor (using the CAM in the manner described by Step 5). If so, go to Step 7. Else, repeat Step 6.

Step 7: Transfer the robot to the location of an unvisited node as per priority ($E > W > N > S$). Infer new location of the robot based on step size, previous coordinates and direction of motion. Store the new grid point coordinates in *Visited grid point_x memory* and *Visited grid point_y memory*. Go to step 2.

Step 8: Stop.

The operation of the algorithm is described now via an example. Consider Figure 4.1. The robot starts at $(40, 40)$ and advances until $(440,40)$ since the priority is along E. Thereafter, since (node along) W has already been visited while further movement along E is prevented by an obstacle, the robot moves in the direction

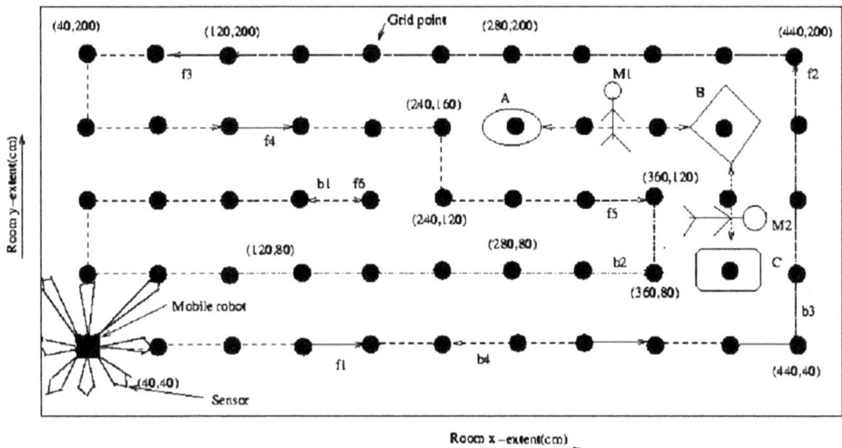

Fig. 4.1. Illustration for the Algorithm (prefix 'f' denotes forward movement while 'b' denotes movement backward towards start node)

that has third priority, namely N and goes up to $(440, 200)$ stopping at each grid point for D_m/s_{min} and taking sensor readings. At $(440, 200)$, the robot advances west until $(40, 200)$, then it moves south until $(40, 160)$. At $(40, 160)$, it moves east to $(240, 160)$ instead of proceeding further south since E has higher priority and further, node $(80, 160)$ has not been visited (yet). The movement east continues until $(240, 160)$. At $(240, 160)$, the only direction that contains an unvisited or unblocked node is south so the robot heads to $(240,120)$. At $(240,120)$, the robot takes a turn east using priority for east (instead of moving further south to $(240, 80)$). Then it moves to $(360,120)$ and so on until it reaches $(200,120)$. At $(200,120)$ it begins to backtrack using *parent node* information at each stage. For this example, the backtracking results in the same path (same set of edges as for the forward motion) getting traced until the root node is reached.

Remark 4.1. The repeated sampling in Step 2 is crucial for determining temporarily occupied nodes (due to dynamic obstacles). Further, the diagonally placed sensors are at the same level as the others so for instance, if there is an obstacle O occupying the NE or SE grid point and the robot is due to move east, one of the diagonally-placed sensors will come into contact with O if O is within d. Also, if the current node's x-coordinate is 40 and we detect (using ultrasonic sensor readings) an obstacle 30 cm away (distance sensor will record 0 since minimum it can handle is 40) from the present node, then node with x-coordinate of 80 is declared as temporarily occupied. However, if an obstacle is 70 cm away from the present node, the next node with x-coordinate of 80 is declared free (the interest in this algorithm is only on declaring the adjacent node as free or occupied).

In step 4, one or more of $N_i, i = 1, \cdots 3$ may be set to a predefined value (such as FF) based on the presence of the boundary of the environment near a node.

In step 5, we determine if there is some *unvisited* neighbor we can move to from the present node or we need to backtrack. In particular, if((visit_tempoccupied[i] $== 0$) and ($N_i \neq FF$)) then there exists a node n_n such that (i) n_n has not been visited (ii) n_n has not been occupied by chairs/humans etc. and (iii) n_n is not outside the boundary (N_i is not FF).

It is worth noting that there is no explicit structure that stores "inaccessible" grid points (corresponding to those that are occupied by static obstacles) - However, a record of temporarily occupied nodes in CAM is essential to prevent robot from trying to "visit" them when it approaches one of those nodes from another direction and thereby make a wrong entry into Visited grid point_x memory (as well as Visited grid point_y memory). Each component of *visit_tempoccupied* vector used in Step 5 can be viewed as a flag. When a flag is set to 1, it corresponds to a node that is temporarily occupied or visited.

Complexity Analysis

Lemma 4.2. *Algorithm Explore_Dynamic_Environment takes linear time for completion of exploration.*

Proof. The exploration process constructs the underlying structure (a graph). The complexity result can be then established as follows. (i) There are only $O(G)$ nodes altogether. (ii) Since each node is connected to at most four other nodes, total number of edges is also $O(G)$. (iii) Each edge is traversed only a constant number of times before the robot returns to the start position. Each node is visited at most four times. Exploration of neighbors of every child node is completed before proceeding to the parent. Memory blocks *Visited grid point_x memory* and *Visited grid point_y memory* are populated during exploration so there is no extra effort. The overall complexity is therefore $O(G)$. *Q.E.D.*

The correctness of the algorithm follows from these facts: The proposed algorithm is an adaptation of depth first search taking priority assigned to various directions for visit of nodes. The assumption that there is a connected set of free nodes (not occupied by a static or dynamic obstacle) enables advancing the robot at each stage. The sensor values enable identification of the P accessible nodes and the termination for the algorithm is determined by the stack contents.

Remark 4.3. The algorithm does not explicitly use G for termination detection since the environment contains obstacles and not all of the G grid points may be accessible. Some of the G points may be in another connected component and if they happen to be in free space, exploration of this connected component has to be carried out separately by first transferring the robot to an accessible node in this component. The same procedure is repeated for all the connected components.

It is worth noting that while some of the sensor values at the next grid node may be inferred from the sensor value at the present node, this increases the amount of hardware for implementation (an explicit subtractor/adder is required for obtaining current sensor value from the previous value). Further, since modules for each sensor operate in parallel, time for distance computation remains the same and is that required for one sensor.

Remark 4.4. The efficiency of the algorithm with respect to the number of steps as expressed by Lemma 4.2 has a direct impact on the energy consumption. In particular, minimizing revisits to nodes saves energy.

Remark 4.5. It is worth noting that in the process of exploration, we are interested in extraction of the underlying structure, namely the graph. One can then make the following observations. The traversal resembles an *Eulerian tour* [71] on a graph: An Euler tour of a graph is a cycle that traverses each edge exactly once, although it may visit a vertex more than once. An Euler tour as such applies to directed graphs but an undirected graph (resulting from the final map in our case) can be converted into a directed one by replacing any edge *ab* in an undirected graph by two directed edges **ab** and **ba**. Considering the "final" graph resulting from our exploration, it can be observed that all the G nodes would have been visited (accessible) if there had been no obstacles. Further, we go back on a given edge during the backtracking. In general, not all nodes in the original set are typically accessible due to the presence of one or more obstacles.

4.4 The Proposed Architecture for FPGA-based Processing

The overall architecture for the proposed algorithm is shown in Figure 4.2. The control unit holds the four registers denoted by N_i for storage of indices of the four neighbors of a given node. We now describe the various elements in the architecture.

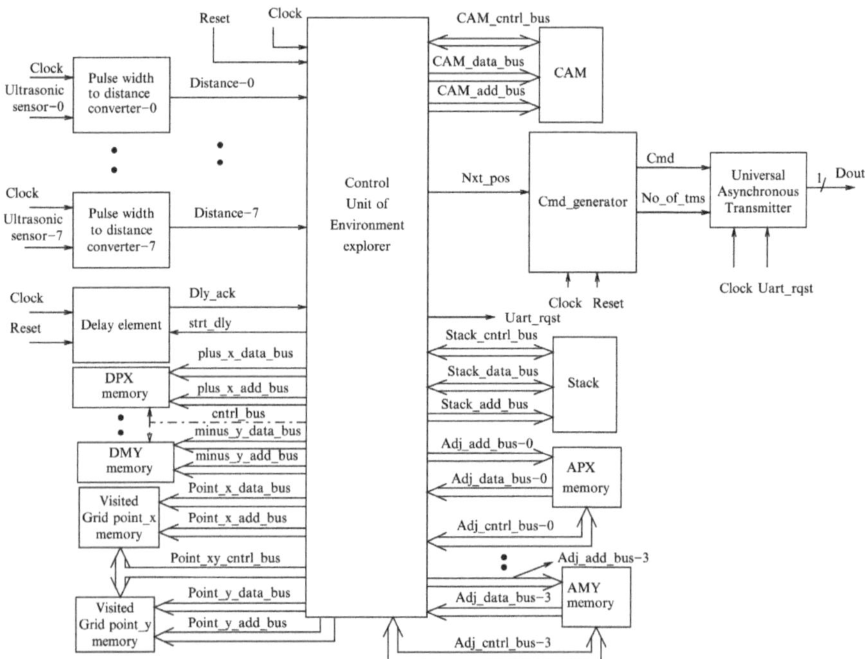

Fig. 4.2. Overall Architecture of Proposed Algorithm

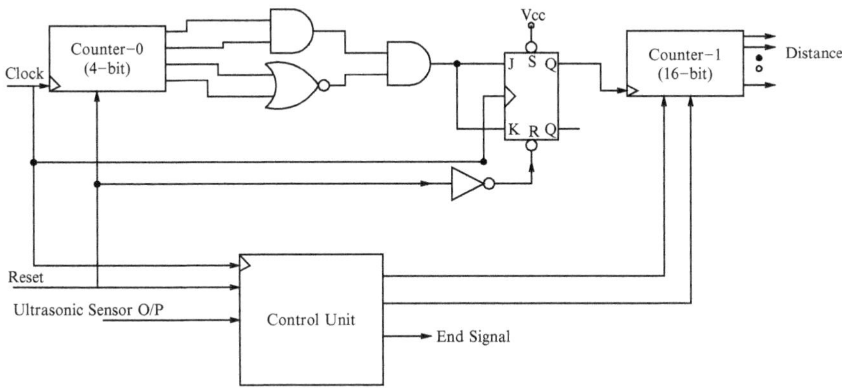

Fig. 4.3. Distance measurement circuit

4.4.1 Pulse Width to Distance Converters

The architecture also consists of eight pulse width to distance converters, one for each ultrasonic sensor. These are realized on the FPGA and the schematic of the converter is depicted in Figure 4.3. Pulse width to distance conversion is accomplished without an explicit analog to digital converter. Further, the design does not use operations such as division expensive in hardware and instead realizes the conversion using shifts.

4.4.2 Content Addressable Memory

Conventional RAM implementations use addresses. An address-based search for a data element is typically performed sequentially and is *slow*. The time required to find an item in memory can be reduced considerably if stored data can be accessed by the content itself rather than by an address [72], [73], [74], [75].

High speed matching is therefore accomplished with a circuit that uses a few registers and a select logic which performs the function of a CAM. The CAM circuit used in our design for storing 16 bytes of data is shown in Figure 4.4.

Eight CAM blocks (of the type shown in Figure 4.4) are used to realize a 128 byte CAM as depicted in Figure 4.5. It is worth noting that our priority search strategy is for an environment with 128 grid points but larger environments can be readily handled.

For the exploration task, the CAM is used to determine whether a node has been visited or not.

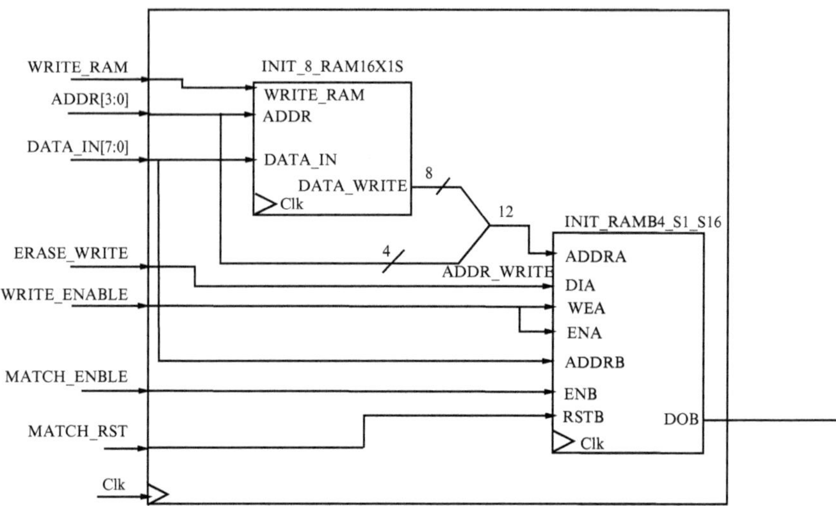

Fig. 4.4. Architecture of CAM for storing 16 bytes of data (Our design is based on data from www.xilinx.com)

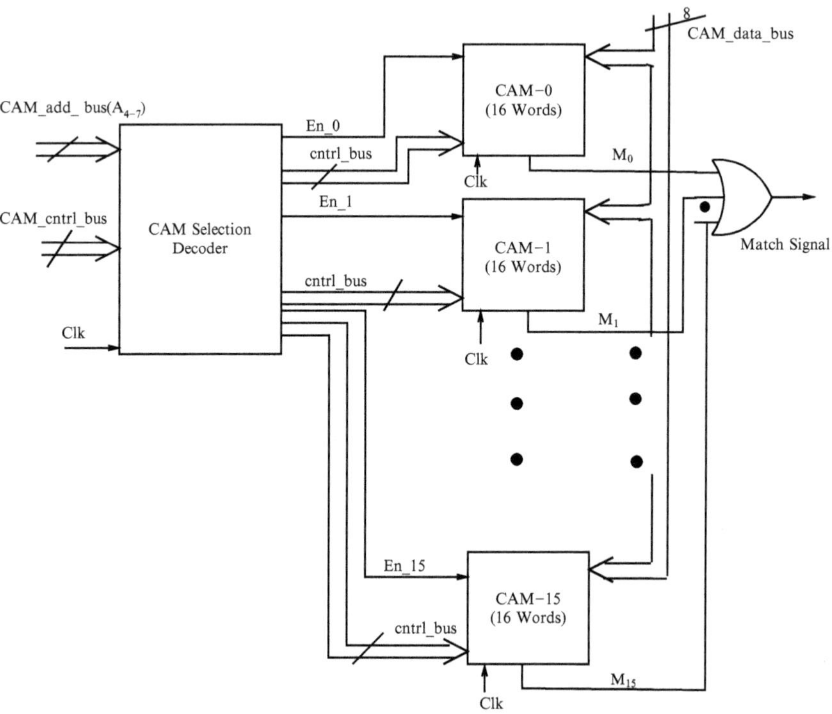

Fig. 4.5. Architecture of CAM for storing 128 bytes

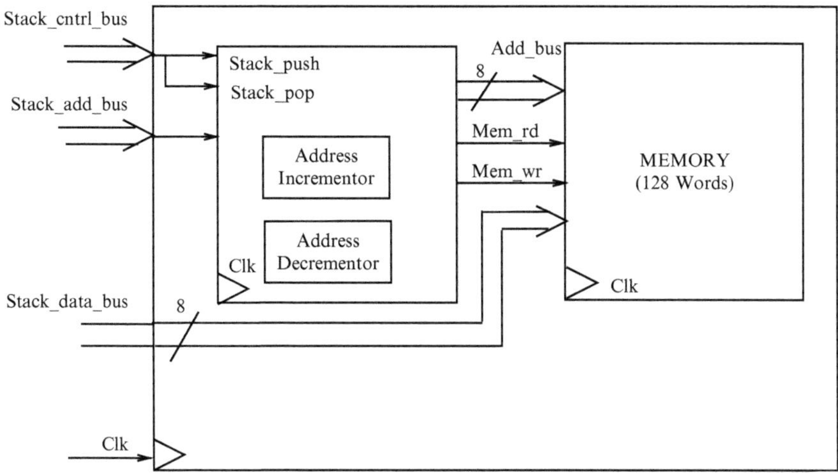

Fig. 4.6. Architecture of stack

4.4.3 Stack Memory

The architecture employs a stack memory for tracking. It is used to store the sequence of nodes visited during the exploration. The stack memory will become empty when the exploration task gets over.

The architecture of the stack is shown in Figure 4.6. The basic elements correspond to incrementing and decrementing addresses besides a memory block.

4.4.4 Universal Asynchronous Transmitter (UAT)

This is for sending commands to the stepper motor interface. The architecture (on which our Verilog program is based) is depicted in Figure 4.7. In Figure 4.7, the UAT starts sending commands serially to the stepper motor control when Reset goes to logic 1. The commands to be sent are stored in advance in the block labelled *Memory* by the main controller (*Control Unit of Environment Explorer* shown in Figure 4.2).

4.4.5 Delay Element

This element in the architecture corresponds to a counter. It is used during exploration task to introduce a delay between the

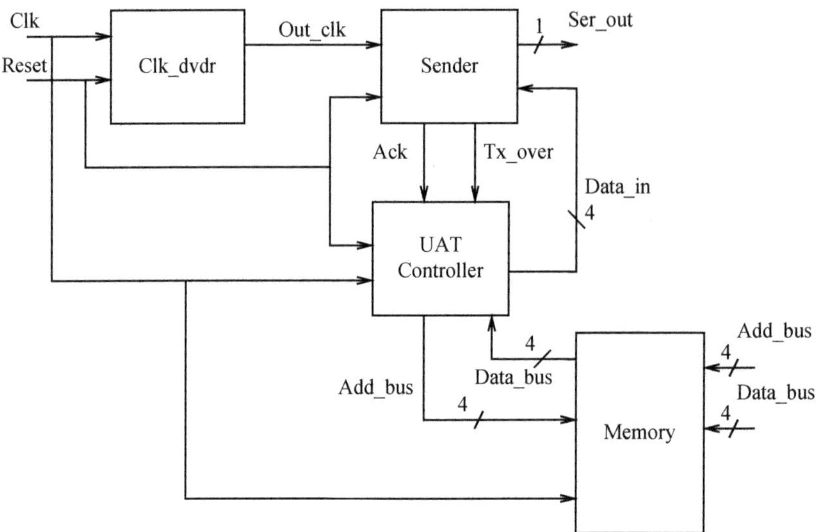

Fig. 4.7. Architecture of Universal Asynchronous Transmitter (UAT)

successive distance measurements to the nearest obstacles. This ensure that distances are measured correctly using the ultrasonic range finders.

4.4.6 Adjacency Information Storing Memory Blocks: APX, APY, AMX and AMY

These memory blocks consist of already stored list of adjacent nodes to each and every node in the N, S, E and W directions as per the grid structure of the indoor environment. During the exploration task, these memory blocks are accessed to test whether the adjacent nodes have been already visited or not.

4.4.7 Memory Blocks Used for Map Construction: DPX, DPY, DMX, DMY, Visited Grid point_x and Grid point_y

These memory blocks are used to store the measured distance values along N, S, E and W directions and X and Y co-ordinates of a node. During the exploration task, at each accessible node, distances to the nearest obstacles in N, S, E and W directions are obtained using ultrasonic range finders. These values are stored in DPX, DPY, DMX and DMY memory blocks. The X and Y co-ordinates of each accessible node are obtained using dead reckoning. These are stored in Visited Grid point_x and Grid point_y memory blocks.

4.4.8 Input Gating for Reducing Energy Consumption

Throughout the design, we place transparent latches with an enable input [76] at the inputs of those parts of the circuit that can be selectively turned off.

Since the FPGA is the processing element on the robot, the device life is of interest. Minimizing energy consumption of the FPGA is valuable since it has an impact on temperature rise due to heating. Increase in temperature in turn contributes to increase in the leakage current [77]. We therefore study ways to reduce energy consumption.

Our contribution is primarily in the identification of the appropriate modules in the robotic mapping application for shutdown when computation is not being performed. The details of application of shutdown to one of the modules in our architecture are now given along with a figure illustrating the shutdown. Consider the Content Addressable Memory (CAM) module depicted in Figure 4.2. The selective shutdown approach as applied to the CAM is illustrated in Figure 4.8. With respect to the proposed algorithm, it is evident that the CAM needs to be active only in Steps 1, 2, 3, 5 and 6 corresponding to initialization, storage of grid point index, insertion of indices of temporarily occupied nodes and determination of node to be visited respectively. Therefore, the CAM is shutdown during the remaining phases of the algorithm.

The selective shutdown mechanism is also applied to other modules in the architecture as follows: The eight pulse width to distance converter modules are active only in Steps 1, 2 and 4 of the algorithm so they are disabled during the remaining period. Further, the memory blocks labelled APX, AMX, APY and AMY are enabled only in Step 4 while the memory blocks labelled DPX, DMX, DPY and DMY are also enabled only in Step 4.

The architecture has also been designed to exploit device-specific characteristics as follows: it is well-known [78], [79] that choice of the appropriate bindings can have a tremendous impact

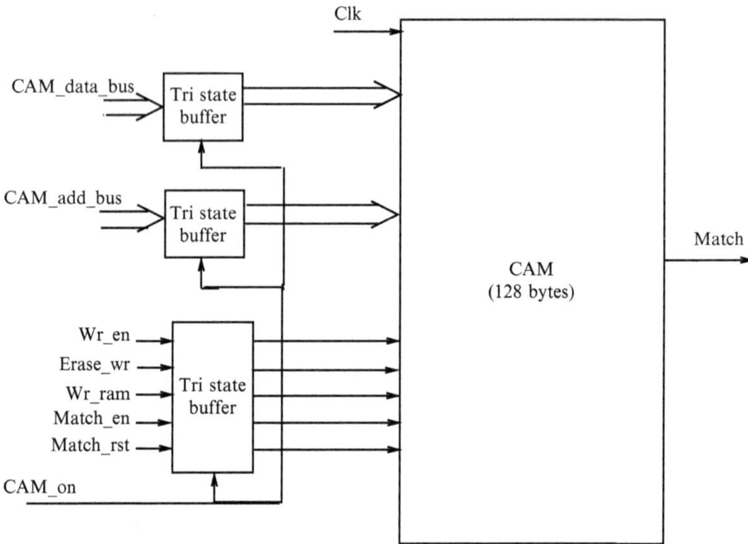

Fig. 4.8. Selective shutdown of CAM for energy-efficient solution

on the power dissipation. In particular, for large storage elements, block RAMs show an advantage in power dissipation over other elements (such as distributed RAMs).

In our robotic mapping application, the storage of data at various grid points accessible to the robot is a crucial step. A variety of memory elements are used as shown in the overall architecture (Figure 4.2). All of these are designed to take exploit the available block RAMs on the FPGA device. Exact usage of the FPGA as well as information on energy consumption are presented in the next section.

4.5 Experimental Results

The hardware design has been coded in VERILOG-2001 and simulated using *ModelSim*. It has then been synthesized using *Synplify 7.6*. The design has been mapped onto Xilinx *XC2S200E* device on a Digilab 2E board. The specifications of XC2S200E device are given in Table 4.1. The floorplan for the implementation of the exploration scheme on XC2S200E is shown in Figure 4.9.

Table 4.1. Specifications of Target FPGA Device

Device	CLBs	Block RAMs	Available user I/O
XC2S200E-PQ208-6	28 x 42	14 (each 4K bits)	146

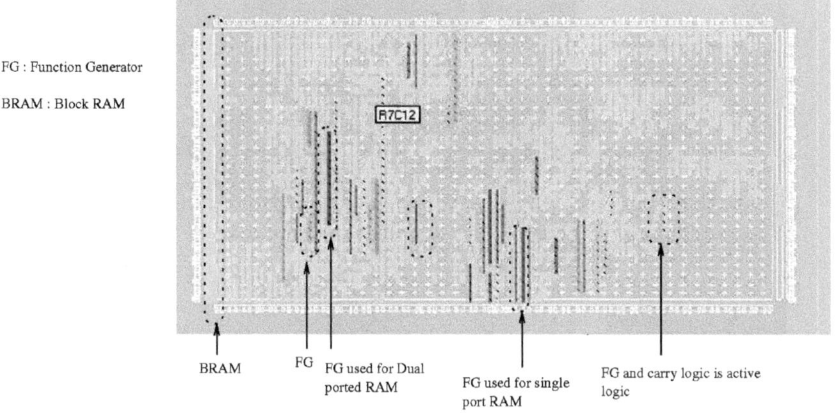

Fig. 4.9. Floorplan for implementation of exploration algorithm on XC2S200E

The first experiment consists of two dynamic obstacles and three static obstacles. It is depicted in Figures 4.10 and 4.11. The mapping experiment is in an area of approximately $4m \times 2m$. The environment consists of 55 nodes ($G = 55$). The number of accessible nodes (P) here is 49. The total number of node visits is 96 (in view of two terminal nodes, one to the left of object B and one right below C). The sensors pointing along W and N will detect moving objects $d1$ and $d2$ *simultaneously* while the robot has reached the node q in Figure 4.11. It may be noted that $d1$ and $d2$ cross the "zone" of the sensors only once during the D_m/s_{min} period :it is clearly this type of a situation that a general-purpose processor (on a PC typically connected to a robot) will, in general, not be able to handle (particularly in the absence of considerable additional hardware such as buffers). The hardware on the FPGA operating *in parallel* determines that the next safe node for the robot is the one located south labelled r.

The results in Table 4.2 can be interpreted as follows. From the specifications in Table 4.1, it can be seen that XC2S200E has a CLB array of 28×42. It is to be noted that every CLB has

Fig. 4.10. Experiment-1 - Robot's sensors pointing along W and N detect two moving humans

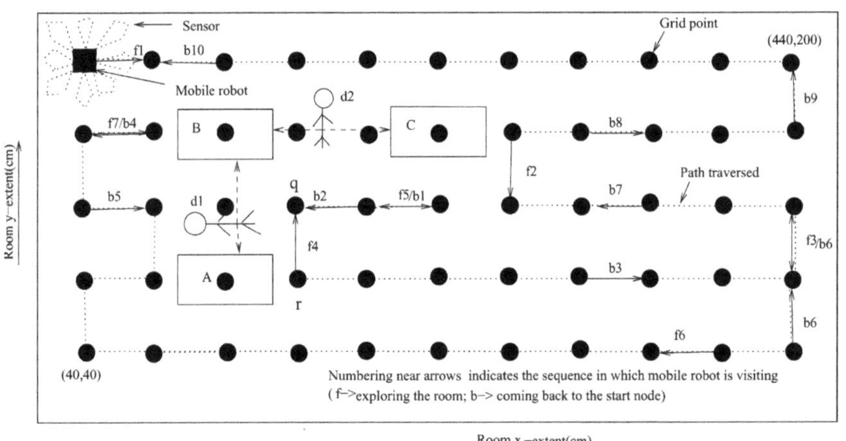

Fig. 4.11. Experiment-1 - Nodes visited based on robot's obstacle detection

2 slices and every slice has 2 LUTs for a total of 4704 LUTs so the usage is 13.5%. Other entries in the table can be interpreted similarly. The area overhead when hardware for selective shutdown is incorporated into the design is very small.

Energy calculations use information on power consumption measured using *XPower*. A sample entry in Table 4.2 is obtained

Table 4.2. FPGA Utilization and Energy Consumption Data for Experiment-1 (G = 55 nodes)

Scenario	Freq (MHz)	LUTs	Block RAMs	Bonded IOBs	Energy Consumption per node (nano Joules)
With BRAMs and selective shutdown	20	639 (13.5%)	4 (28%)	28 (19%)	28.75
With BRAMs and no shutdown	20	621 (13.2%)	4 (28%)	28 (19%)	46.2
With selective shutdown and deselection of BRAMs	20	1143 (24.2%)	0	28	49
Deselecting BRAMs and no shutdown	20	1125 (23.9%)	0	28	75

as follows. In row 1, the energy per node is obtained using power consumption computed to be 23 mW, the total number of clock cycles (25) and the operating frequency of 20 MHz. The operating frequency is chosen based on the ability of the FPGA to successfully (and consistently) input data from the sensors. It may also be noted from Table 4.2 that selection of BRAMs contributes to energy savings: this can be attributed to the lumped nature.

The second experiment for also consists of two moving objects but the static objects are different. In particular, one of the static objects (denoted by C) in Figure 4.13 is a desktop computer as shown in Figure 4.12. Once again altogether 55 nodes ($G = 55$) are present. The results of mapping are as follows. The nodes visited for experiment 2 are shown in Figure 4.13. P here is however less than that for Experiment-1 and is only 47. The total number of node visits is 92 (there are 3 terminal nodes visited just once while there are two nodes denoted by s and t that are visited three times). The robot determines at q the appropriate node to visit next. In particular, the robot's sensors pointing along E and N *simultaneously* detect moving objects $d1$ and $d2$ respectively and the FPGA processes the sensor data to determine that the next safe node is r. The complete map is shown in Figure 4.13. The resource utilization on the FPGA remains the same as for Experiment 1. However, there are some differences with respect to the

Fig. 4.12. Experiment-2 - Robot's sensors pointing along E and N detect the two moving humans

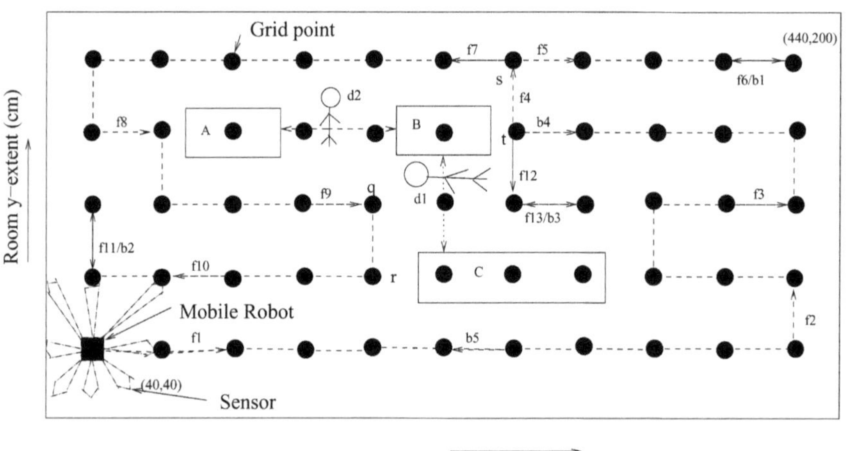

Fig. 4.13. Experiment-2 - Nodes visited; (prefix 'f' denotes forward movement while 'b' denotes movement backward towards start node)

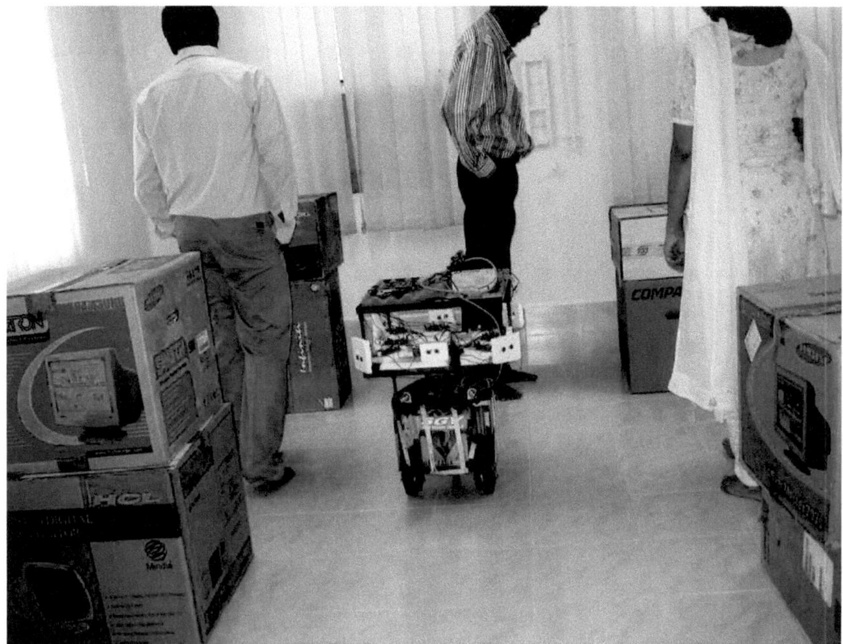

Fig. 4.14. Experiment-3 - Robot's sensors pointing along E, N and W detect three moving humans

energy consumed since P here is different as also the total number of nodes visited.

We now consider an experiment with three moving objects. It is depicted in Figure 4.14. As in the previous experiment, altogether 55 nodes are present. The results of mapping are as follows. The nodes visited for experiment-3 are shown in Figure 4.15. P here is 48. The total number of node visits is 94. When the robot is at node q in Figure 4.15, it observes humans moving along E, W and N. The hardware on the FPGA operating *in parallel* determines that the next safe node for the robot is the one located south so the robot traces back the path it came on. The resource utilization on the FPGA remains the same as for Experiment-1.

A variation of the proposed algorithm has been developed to work for a purely static environment. The variation has been tested on environments with multiple static objects. We begin with an experiment consisting of four static objects (Figure 4.16). The mapping area is approximately $4m \times 3m$. The nodes visited and the map constructed are shown in Figure 4.17. G here is 82. The

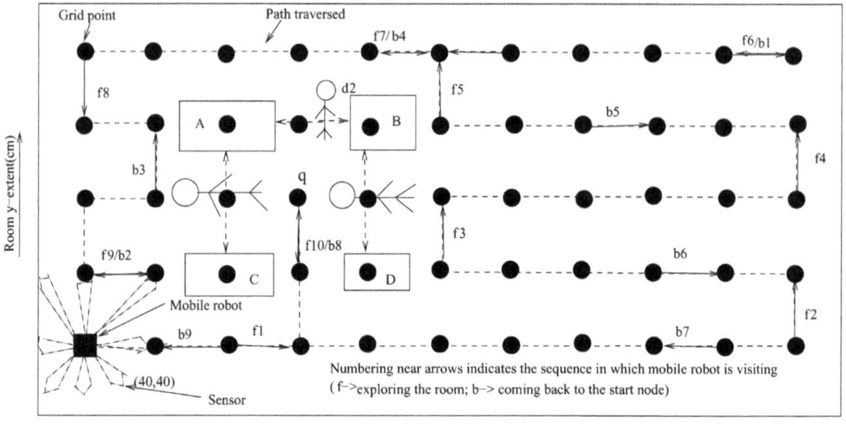

Fig. 4.15. Experiment-3 - Nodes visited based on robot's obstacle detection

Fig. 4.16. Experiment-4 - Robot's sensors detect inaccessible corner due to two static obstacles

results of mapping task are as follows. The number of accessible nodes P is only 67. It is worth noting that accessibility is influenced by (i) obstacles occupying some grid points and (ii) obstacles preventing access of some grid points keeping in view the

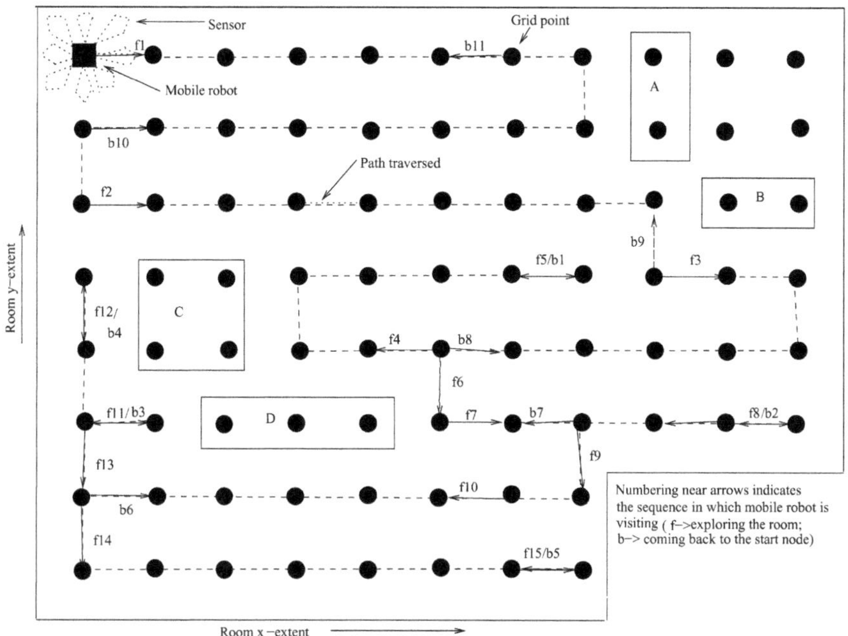

Fig. 4.17. Experiment-4 - Nodes visited

Table 4.3. FPGA Utilization and Energy Consumption Data for Experiment-4 (G = 82 nodes) - Static Objects

Scenario	Frequency (MHz)	LUTs	Block RAMs	Bonded IOBs	Energy Consumption per node (nano Joules)
With BRAMs and selective shutdown	20	707 (15%)	8 (56%)	28 (19%)	33.75
With BRAMs and no shutdown	20	690 (14.6%)	8 (56%)	28 (19%)	55
With selective shutdown and deselection of BRAMs	20	1715 (36.4%)	0	28	65
Deselecting BRAMs and and no shutdown	20	1706 (36%)	0	28	105

size of the robot and the clearance available to pass through the gap between disjoint obstacles. In particular, nodes to the "right" of A and "above" B are also inaccessible. The total number of node visits is 132. The resource utilization results are presented in Table 4.3. The slight increase in the number of LUTs in Table 4.3

Fig. 4.18. Experiment-5 - Robot moving in the presence of purely static obstacles

when compared to the results in Table 4.2 can be attributed to increased storage for the 82 nodes (as compared to that required for 55 nodes). As before, the advantages of selecting BRAM and enforcing the selective shutdown are seen in terms of reduction in energy consumption. The increase in the number of LUTs when selective shutdown is enforced turns out to be marginal. The energy consumption proportionally grows as a function of the number of nodes visited.

We now present the results of exploration applying the algorithm for another environment with static obstacles. Here $G = 40$. Figure 4.18 shows the objects and the robot. The nodes visited are shown in Figure 4.19. The mapping experiment is in an area of approximately $3m \times 2m$. There are altogether 32 accessible points. The resource utilization results are presented in Table 4.4. The data is the same as that for Table 4.2 since memory allocation for 40 and 55 nodes is based on requirements for a node number

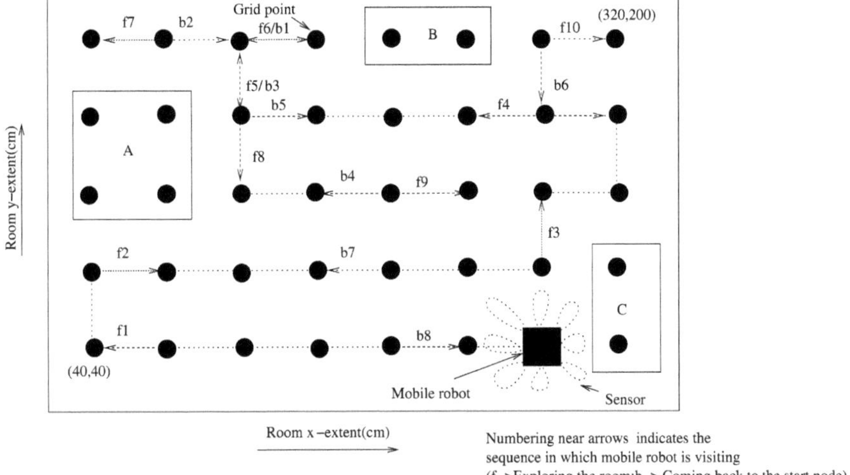

Fig. 4.19. Experiment-5 - Nodes visited

Table 4.4. FPGA Utilization and Energy Consumption Data for Experiment-5 (G = 40 nodes)

Scenario	Freq (MHz)	LUTs	Block RAMs	Bonded IOBs	Energy Consumption per node (nano Joules)
With BRAMs and selective shutdown	20	639 (13.5%)	4 (28%)	28 (19%)	28.75
With BRAMs and no shutdown	20	621 (13.2%)	4 (28%)	28 (19%)	46.2
With selective shutdown and deselection of BRAMs	20	1143 (24.2%)	0	28	49
Deselecting BRAMs and no shutdown	20	1125 (23.9%)	0	28	75

corresponding to a power of 2 (in this case 64) that is greater than 40. The data for energy consumption per node is the same as that for $G = 55$.

4.6 General Remarks about Code and Demonstration

Appendix A presents the main Verilog modules for the FPGA-based robotic exploration. The complete set of programs is available from

http://www.ee.iitm.ac.in/~sridhara/Verilog_ExploreAlg.

Video files showing the implementation of the algorithm on our robot are available from the URL `http://www.ee.iitm.ac.in/~sridhara/video_dyn_map`. The video can be seen using Windows Media Player or RealPlayer.

A brief introduction to the Verilog programs (in Appendix A) is given here. The top level module $explore\_env\_using\_dfs\_128$ is the core of the hardware design. The module is a synchronous module and has clk as an input signal. It also has $start\_exp$ as an input signal to start exploration. Further, $reset1$ is an input signal to initialize its internal registers as well as registers of instantiated modules. $sen\_out\_cen\_plus\_x$, $sen\_out\_rght\_plus\_x$ etc. are input signals for processing the ultrasonic range finder outputs. Output signals such as cam_on, $stack\_on$, co_store_on and cmd_gvn0 are used for facilitating the testing of CAM, Stack, Coordinates storing memory and command generating modules respectively. Output signals such as an, ssg and ldg are used to enable LEDs and two seven segment displays of the FPGA board for verifying the functionality of the instantiated modules. bin_7seg, $seven\_seg$, dyn_dfs_128, $uart$, $ultra\_sonic\_sensor$, $delay$ and $more\_delay$ are the various modules instantiated in the module named $explore\_env\_using\_dfs\_128$. The bin_7seg module works as BCD to 7-segment decoder while the $seven\_seg$ module is a control unit which enables four seven segment displays on the FPGA board to display a 16-bit (signal) value (alternatively, the value of two 8-bit signals or the value of four 4-bit signals). The dyn_dfs_128 is a control unit of the main module. The control unit is implemented using a state machine which has 40 states. More details of the module are given in the next paragraph. The $uart$ module is used for sending commands to stepper motor driver using serial communication. The data rate of the serial communication is 10 Kbps. The $ultra\_sonic\_sensor$ module converts a pulse produced by the ultrasonic range finder into an equivalent 16-bit number. The $delay$ module is used to push the control unit into idle state during distance measurement to the nearest obstacles. The $more\_delay$ module is used to push the control unit into idle state when robot is moving from one node to the next.

The dyn_dfs_128 module instantiates CAM_RAMB_128, $adja\_nodes\_store0\_128$, $adja\_nodes\_store1\_128$, $adja\_nodes\_store2\_128$, *adja_nodes_store*3_128, *visted_node_store_*128, *node_x_cord_store_*128, *node_y_cord_store_*128, *stack_*128 and *cmd_gen_*128 modules. The *CAM_RAMB_*128 module is used to store the index value of visited node. The *adja_nodes_store*0_128, *adja_nodes_store*1_128, *adja_nodes_store*2_128 and *adja_nodes_store*3_128 modules provide the indices of nodes in four directions to the node currently visited by the mobile robot. The visited node index is stored in *visted_node_store_*128, while the (x, y) coordinates are stored in *node_x_cord_store_*128, and *node_y_cord_store_*128 modules respectively. The *stack_*128 module is for three purposes. One purpose is to store visited node index, the second purpose is to get the index of a node in the event of backtracking and the third purpose is to stop the exploration. The *cmd_gen_*128 module is used to generate command words, which are sent to the stepper motor driver using serial communication by the *uart* module.

4.7 Conclusions

This chapter has presented a new VLSI-efficient algorithm for robotic exploration of a *dynamic* environment. The approach presented guarantees uniform coverage of the space by the robot. Further, with little additional effort, terrain acquisition can also be performed. A key ingredient of the approach is *parallel processing* of data from multiple ultrasonic sensors to accurately determine information about temporarily/permanently occupied nodes without using external buffers. It is worth noting that if the robot is permitted to move along eight directions (instead of four), the FPGA-based approach can readily handle processing of activity at all the eight neighbors in parallel.

A new architecture for environment exploration is also proposed. Features of the architecture are area and energy-efficient processing of input data. Experimental results with a mobile robot developed in our laboratory and equipped with a low-cost FPGA, and ultrasonic sensors demonstrate the feasibility of using an area-efficient and high-speed alternative to general-purpose processors. It is worth noting that the solution operates without a PC as part of the loop. The approach also does not require any additional hardware (such as buffers) or expensive range finders. Since FPGAs allow simple pin-level control, a large number of sensors can be readily handled *in parallel* using user I/O pins. Further, a key feature of our FPGA implementation is small usage of the device components for a fairly large environment.

5

Hardware-Efficient Landmark Determination

5.1 Motivation for Landmark Determination

In the previous chapter, the scheme for exploration generated a subset P (of the G grid points) that the robot can reach and obtain sensor readings. A map indicating connectivity of "adjacent" nodes (for example, a node at (40,40) and (80,40) in Figure 4.11) was also generated.

In general, for any point $p \in P$, there may be many points in the set P that are accessible based on sensor values. Depending on sensor data stored at a pair of "non-adjacent" sample points, such as (40,40) and (120,80), one can determine if (40,40) and (120,80) can be connected. However, this would result in a very large number of interconnections (effectively edges) leading to a graph that is cumbersome to deal with for shortest path calculations and related computations. This chapter therefore examines the possibility of generating a subset of P that would be appropriate for navigation. The procedure for generation of the subset is presented in the paragraphs to follow. It turns out that values of sensors pointing along $+x, -x, y$ and $-y$ are adequate to generate this information.

Typically, among P, there may be many points in close proximity and it is desirable to simply have representatives of sets of nearby points. As pointed out, good representatives are appropriate for navigation purposes since they enable rapid path finding (in other words, it will take less time to compute the shortest path on a smaller graph). We call these representatives as *perceptual landmarks* or merely as *landmarks*.

K. Sridharan and P. Rajesh Kumar: *Robotic Exploration and Landmark Determination*, Studies in Computational Intelligence (SCI) **81**, 63–86 (2008)
www.springerlink.com   © Springer-Verlag Berlin Heidelberg 2008

Our primary objective here is development of a new hardware-efficient algorithm for construction of landmarks. Parts of this are reported in [80]. In particular, a space-efficient scheme is valuable since the exploration task takes place using the FPGA on the mobile robot (described in the earlier chapters). Further, we develop a scheme that employs multiple processing elements since it is appropriate to rapidly devise paths in a dynamic environment. Some assumptions and terminology required for the development of the algorithm are first presented.

5.2 Assumptions and Terminology

We build landmarks based on sensor information (at accessible points) along four directions only. This is keeping in view the four directions of motion for the robot.

As already mentioned, the robot attempts to visit a set of grid points. At any accessible point, data get stored in a *sample point* structure and would be given by $[W_0[s], \cdots, W_5[s]]$. P sample points are assumed to be available altogether.

The notion of a *place unit* which is a structure with six components is used in this chapter. The first two components of a place unit contain coordinates (defining the location of a place unit) while the remaining four correspond to sensor values along $x, -x, y$ and $-y$ directions. A set of K place units (each of type *place unit* and denoted by $N[i], i = 1, \cdots, K$), will represent perceptual landmarks (or simply *landmarks*) after training and appropriate adjustment. The representation for place unit $N[i]$ will be denoted by $[N_0[i], \cdots, N_5[i]]$. The place unit components are randomly initialized. Figures 5.3 and 5.4 illustrate the notions of sample point, place unit and landmark.

We obtain landmarks from the place units. Clustering takes place and the values of the components get altered. The activity at each place unit that can be performed *in sequence* on a single general purpose processor (one by one for all the place units) is parallelizable. Hence, each place unit can be mapped and implemented as a separate Processing Element (PE) in hardware. It turns out that a small number of landmarks (therefore the same number of PEs) is usually adequate for indoor lab environments.

For example, 8 to 10 landmarks may be adequate for an area of approximately 5 sq. m (if obstacles are not many). The maximum number of PEs used in the hardware implementation presented in this chapter has been determined based on estimation of the amount of hardware required (and therefore space consumed) in each PE.

The place units are trained: the randomly assigned initial values to each component of a place unit get altered depending on the robot's data at each sample point. Training takes place with respect to each sample point and for several iterations. The place units whose components will have changes are identified by determining a *winner place unit* for each sample point and at each iteration. The procedure is as follows. The distance of each place unit from a given sample point is found using city-block metric (l_1 norm) since it is digital hardware-friendly. The distance in l_1 norm between a place unit $N[i]$ and a sample point s is given by $|N_0[i] - W_0[s]| + |N_1[i] - W_1[s]|$. (It should be noted that training in our scheme uses only the first two components to get a good distribution of place units; more details are presented in Section 5.3.) The nearest place unit (using the l_1 norm) is declared to be the winner place unit. The components of the winner place unit are adjusted using a parameter named *learning rate* and denoted by a. The new values for the components are obtained using Eq. 5.1.

$$N_0^{new}[i] = N_0^{old}[i] + a[W_0[s] - N_0^{old}[i]]$$
$$N_1^{new}[i] = N_1^{old}[i] + a[W_1[s] - N_1^{old}[i]] \tag{5.1}$$

a is assigned a positive value between 0 and 1 initially. When the learning rate is high (>0.5), the network tends to learn only the latter samples while for very small values of a (<0.01), the ability to learn slows down.

In general, besides the winner place unit, a few other place units within some distance of the winning place unit may also have their components altered. These place units are identified by a parameter named *neighborhood radius* (denoted by r). The adjustment of components of these neighbors also follows Eq. 5.1. An initial value for the neighborhood radius is determined from the dimensions of the grid of place units. While the network is trained via a number of iterations, the learning rate does not remain fixed.

In order to generate *good landmarks*, the learning rate has to be steadily decreased and this process begins after a certain number of iterations. One quantity of interest here is the threshold value for iteration number (denoted by I_1 henceforth) from which a is steadily reduced.

5.3 Proposed Algorithm

5.3.1 Key Ideas

Each processing element (PE) has six components and two of them correspond to x and y coordinates - the components are initially assigned random values indicating a random "location" of the place unit in the space in which the robot operates (along with random sensor values at those locations). The place units settle after a training phase followed by an adjustment phase.

The (explored) environment data is stored in global memory. The number of sample points is quite large (in comparison to the number of place units). For our experiments, the total number of iterations is set to 1000. The learning rate a has been initialized to 0.5 for this research based on some experiments. I_1 defined in section 5.2 has been set to 500 since we set the total number of iterations to 1000. In general, a is decreased after approximately half the number of iterations.

The proposed algorithm has three primary steps: In the first step, place units are trained taking only the coordinates of the sample points stored in the memory: the sensor values are not used for training as such since our initial experience using all the components has resulted in a less desirable distribution of place units. The assignment and use of sensor values to determine information on adjacency between place units are described below. The training process identifies a winner place unit for each sample point at each iteration. The components of the winner place unit are adjusted as per Eq. 5.1.

In the second step, each of the place units gets translated to a sample point in the vicinity in the 2D space (this takes into account the situations where a place unit after training goes into the interior of some object - however, since there is no prior knowledge of the object geometries, translation of all place units is essential).

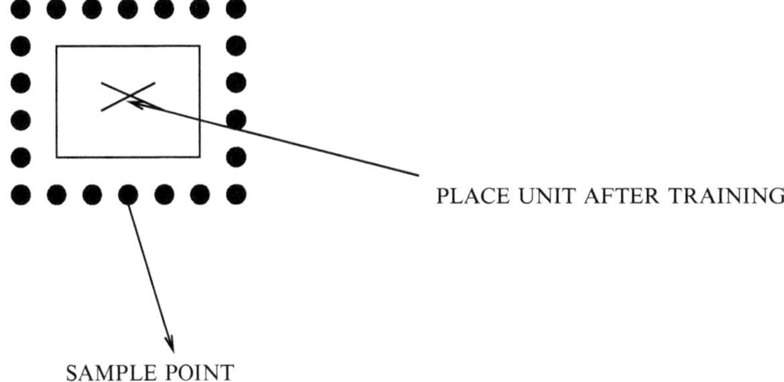

Fig. 5.1. Scenario indicating adjustment of place unit locations

Figure 5.1 provides an illustration for the situation where a place unit could be after training suggesting the need for adjustment of position.

The sample point in the vicinity to which a place unit is translated is determined as follows. Sample points may *not be uniformly spaced* since the location of an obstacle may prevent R from getting data at a specific grid vertex. Place unit locations correspond to one of the sample points after translation. For example, in an environment of size $400cm \times 400cm$ (the environment size here uses information on step size (40 cm) of the robot used), the sample points may have coordinates given by (40,40), (40,120) and so on. If a place unit has coordinates of (37,35) after training, it will get adjusted to (40,40). It should be noted that a grid point such as (40,80) may not be a *valid* one owing to the presence of an obstacle.

The sensor values are assigned to each place unit after the adjustment of location and they precisely correspond to those of the sample point (to which a place unit has translated).

In the third step, the map is generated taking into account the sensor values stored for each place unit after movement. In particular, we build an adjacency structure containing information on the list of place units adjacent to each place unit. Figure 5.2 illustrates how a place unit may be connected to another place unit on the basis of sensor values.

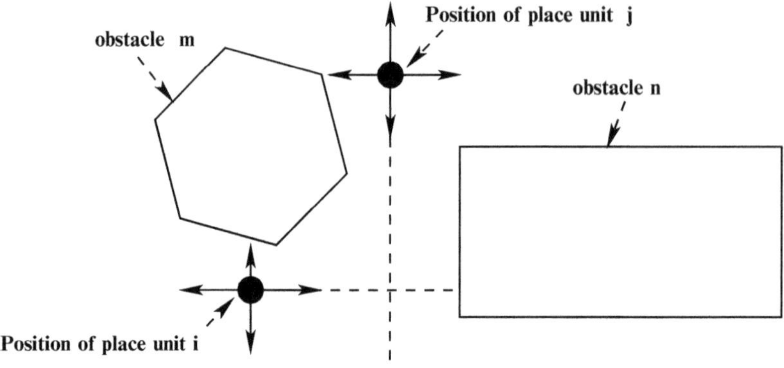

Fig. 5.2. Connecting a place unit to another

5.3.2 The New Algorithm

Algorithm FIND_LANDMARKS_AND_CONNECTIVITY
Input : Sample points of the explored environment
Output: Adjacency structure of size $K \times K$. A place unit $N[i]$ is
connected to another place unit $N[j]$ typically by two segments
(one horizontal segment taking sensor data along $x(-x)$ for $N[i]$
and $N[j]$ and one vertical segment taking sensor data for $N[i]$ and
$N[j]$ along $y(-y)$).

Step 1: Initialize learning rate (a), neighborhood radius (r) and I_1.
Step 2: Initialize first two components of all place units **in parallel**
with random values.
Step 3: for j = 0 to M
        if (j $>= I_1$)
          a = a*0.5;
       for s = 0 to P
          for each of the K place units **do in parallel**
            compute the distance in city-block metric
            (as per Eq. 5.1)
         end
       Determine the winner place unit by binary
       tree-structured comparison.
       Update the components of the winner place unit and
       the four neighbors (along $x, -x, y$
       and $-y$) as per Eq. 5.1 in section 5.2

```
                end /* for sample points */
            end /* for iterations */
```

Step 4: for all place units **do in parallel**
            find a sample point in the vicinity in
            constant time and adjust x and y coordinates
       end
Step 5: for all place units **do in parallel**
            load the remaining components of
            all the place units with the sensor values
            of the vicinity sample point (found in Step 4).
       end
Step 6: for each place unit $N[i]$, $1 \leq i \leq K$
            determine connectivity to place units
            $N[j]$, $j = 1, \cdots, K$, $i \neq j$
            based on sensor values stored;
            repeat for all place units.
       end

A summary of key steps is as follows. In step 3, training begins. In step 4, the locations of the place units are adjusted. In step 5, sensor values are loaded in the place units using those of the sample points (to which the place units are transferred). Step 6 establishes connectivity between the place units. Parallelism in various steps contributes to the desired speedup. Sequential operation in Step 6 ensures that the design fits in a moderately-sized FPGA device.

Remark 5.1. For our experiments, M in the algorithm has been set to 999. In general, when the number of iterations increases, the training results are better. P corresponds to 196. This number has been chosen taking dimensions of a room and the step size for our robot. I_1 has been set to 500 (in accordance with the chosen value for M).

Remark 5.2. We find a sample point in the vicinity of a place unit in constant time (in step 4) by checking between which pair of x values of sample points, the x value for the place unit falls (in constant time). Once the x adjustment is done, the same procedure is repeated for the y value of the place unit.

Analysis of Time Complexity

The time complexity of the proposed algorithm can be analysed as follows. Step 1 and 2 take $O(1)$ time. Step 3 takes $O(MP + MP \log K)$ time since all place units find distance to a point s in parallel and then binary-tree structured comparison takes $O(\log K)$ time to identify the nearest place unit. Steps 4 takes $O(1)$ time and the arguments are as follows: Computation of the nearest sample point for a given place unit is handled by multiple comparators. For example, if a place unit after Step 3 is at an x-value of 30 and the nearest sample point is at 40, the comparisons with others such as those at 80, 120 etc. happen in parallel and only the (comparator) unit corresponding to 40 would output a logic 1. Therefore, for a given place unit, we find the nearest sample point in constant time. Since the process is performed in parallel for all the place units, Step 4 takes $O(1)$ time altogether. Step 5 also takes $O(1)$ time. Step 6 takes $O(K^2)$ time. The overall complexity is therefore $O(MP + MP \log K + K^2)$ or $O(MP \log K + K^2)$. Typically, $K < M$ and further, $K < P$. Hence, the overall complexity can be expressed as $O(MP \log K)$.

Remark 5.3. It is clear that a sequential version of the algorithm presented would have a higher computational complexity. In particular, Step 2 would take $O(K)$ time. Step 3 would take $O(MPK)$ time. Steps 4 and 5 require $O(KP)$ time (since for each place unit, it takes $O(P)$ time in step 4 to output the nearest sample point) while Step 6 takes $O(K^2)$ time. The overall complexity is therefore $O(MPK)$. It should, however, be noted that the KP factor in step 4 and the lack of parallelism in Step 2 (in addition to the lack of parallelism in Step 3) lead to a much larger execution time on a PC. Further, a PC-based solution demands more (physical) space than an FPGA-based solution.

Simulation Results

The sequential version has been simulated on a PC with Pentium IV processor (1700 MHz) and 256 MB RAM to verify correctness. Different types of objects (polygonal, curved) were included in the simulation.

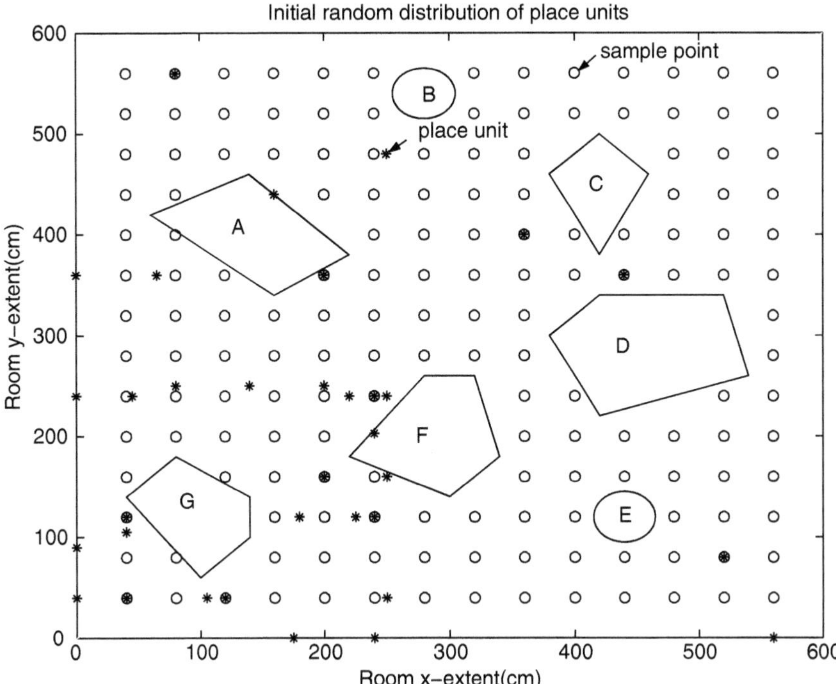

Fig. 5.3. Simulation for a room of size 6m x 6m: Place units before training are shown

The simulation is for a room of size 6m x 6m. Various types of objects have been included. The step size was taken using that of the robot. A total of 162 sample points are present. The number of landmarks is 36. The place units before training are shown in Figure 5.3. Landmarks after training are shown in Figure 5.4. It can be observed that the landmarks cover the space well. Since they represent a small subset of the sample points, computations for path finding would require less time (than that using the entire set of sample points). The map obtained is therefore valuable for path finding.

5.4 The Proposed Architecture

The overall architecture is shown in Figure 5.5. Internal details of various elements are given in the subsections to follow.

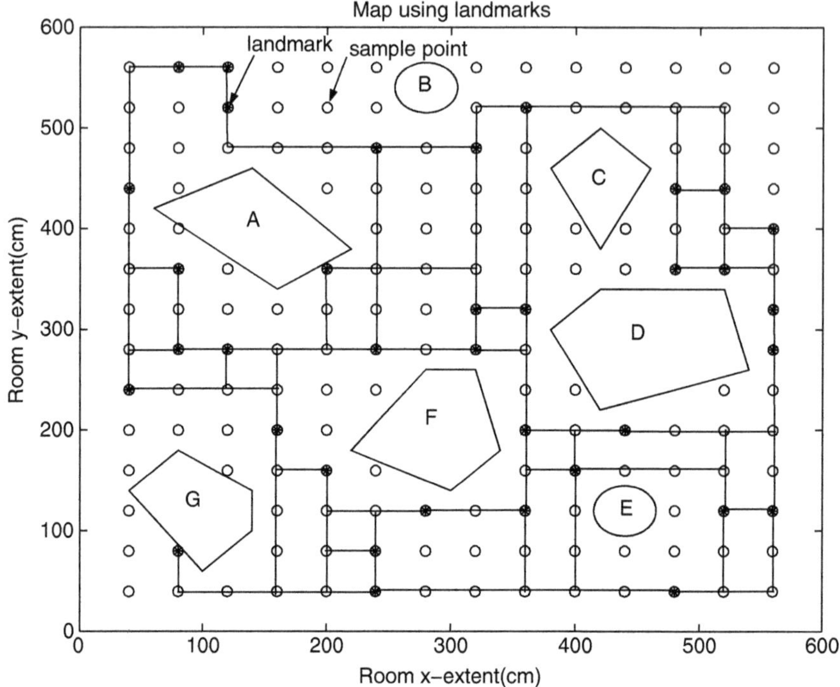

Fig. 5.4. Simulation for a room of size 6m x 6m: Landmarks shown

5.4.1 Random Number Generation

Step 2 of the proposed algorithm assigns random values to the coordinates of the place units. This is accomplished in hardware by means of a random number generator in Figure 5.5.

The architecture of the random number generator is shown in Figure 5.6. The random number generator shown in Figure 5.6 produces a 4-bit address that is used for initialisation of the first two components of all the place units. It is built using a Linear Feedback Shift Register (LFSR) [77]. The LFSR consists of 4-one bit registers. A 4-bit value is applied as a seed to initialise the random number generator. The LFSR then cycles through 15 states before repeating the sequence, so as to produce a pattern of 4-bit random numbers.

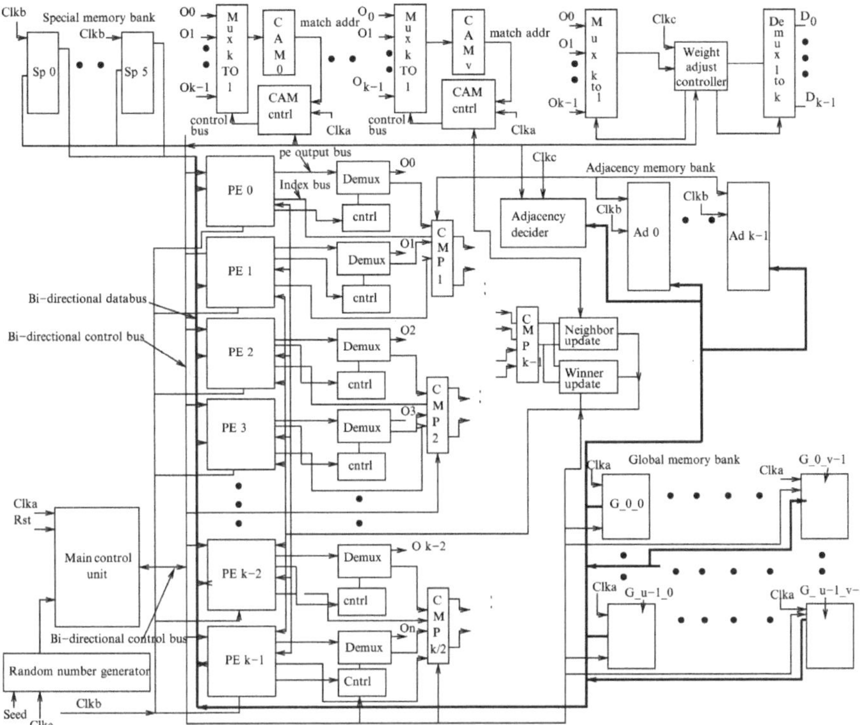

Fig. 5.5. Overall Architecture of Landmark Identifier and Connectivity Finder

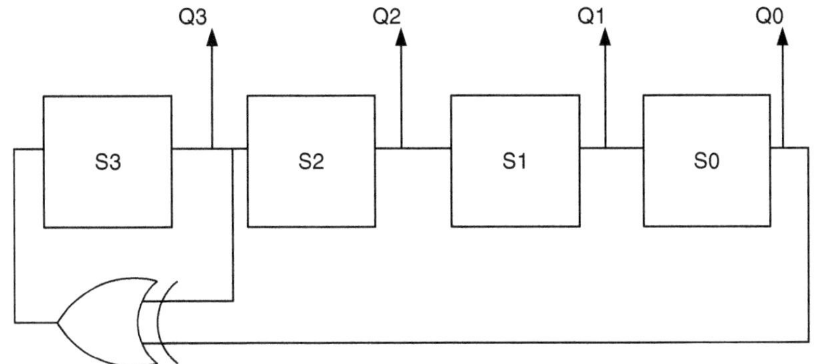

Fig. 5.6. Architecture of the Random Number Generator

5.4.2 Processing Element (PE) Structure

The circuitry for each PE is shown in Figure 5.7. The circuitry for the block labelled "Training Unit" in Figure 5.7 is shown in Figures 5.8 and 5.9.

A feature of the design is that the internal state machine of Figure 5.7 is operated at twice the frequency of the main controller to speed up the training (In the implementation, the clock frequency of the main controller is obtained and the frequency of the internal state machine is set suitably; the critical path in the actual main controller circuit is approximately three times that of the circuit for the internal state machine). The internal state machine sequences the tasks such as initialization of components, computation of difference between components of place units and sample points, loading of sensor values into the corresponding components of the place unit, and loading of elements of a row of the adjacency structure into adjacency memory upon receiving control signals from the main control unit. A set of six 16-bit registers is used to store components of the PE (corresponding to a place unit). Bidirectional data buses are used during initialisation of components, training of the PE and loading of sensor values.

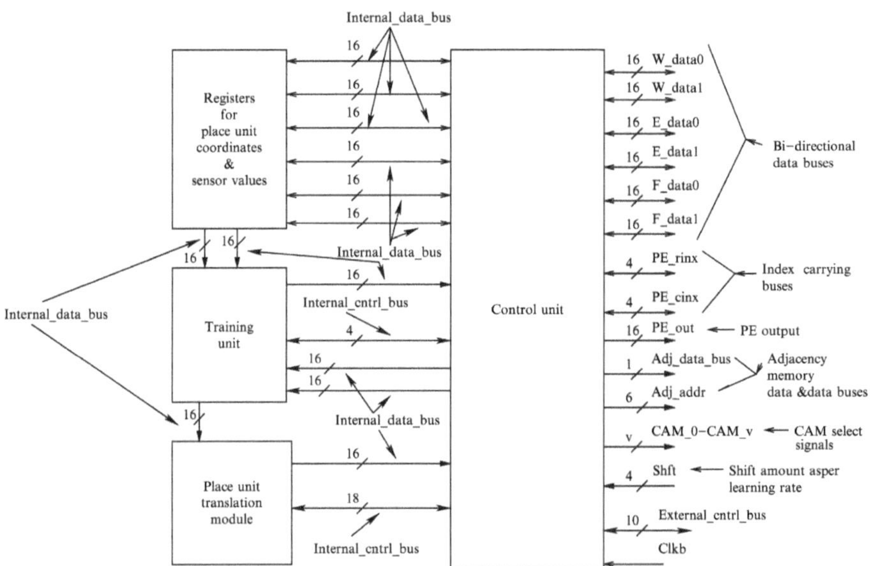

Fig. 5.7. Internal block diagram of a PE

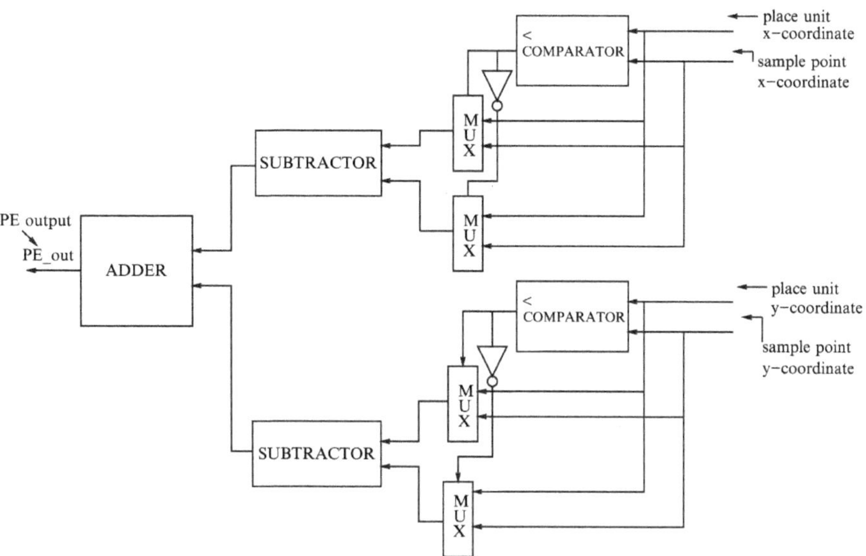

Fig. 5.8. Circuitry for winner determination

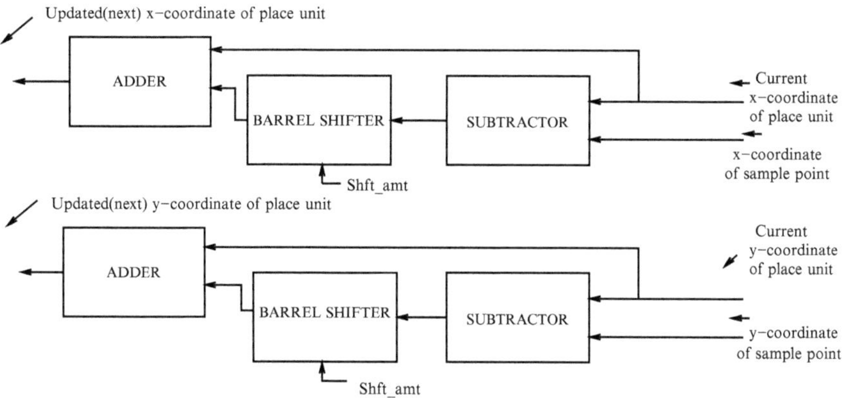

Fig. 5.9. Circuitry for updation of components of winner PE

Index carrying buses supply the identity of the PE to the external winner determination circuit during training.

Another feature of the proposed architecture is the use of *barrel shifters* to speed up updating of the components of place units depending on the value of the learning rate parameter. Two barrel shifters are used to compute two place unit components. The ALU performs computations to measure the distance between first two components of the place unit and coordinates of sample points.

Each PE interacts with global memory (during Steps 3 and 5) whose organization is presented in the next subsection.

5.4.3 Global Memory Organisation

In the proposed algorithm, steps 3 and 5 require access to global memory. The organization of this memory is shown in Figure 5.10. Global memory is organised in 2-D as a set of $u \times v$ memory blocks, where u denotes the maximum number of steps that can be made along x-axis and v denotes the number of components in a place unit. Each memory block holds q 16-bit elements, where q denotes maximum number of steps that can be made along y-axis. When $v = 6$, the place unit holds coordinates of sample point position and distance data from sensors along $+x, +y, -x,$ and $-y$. The

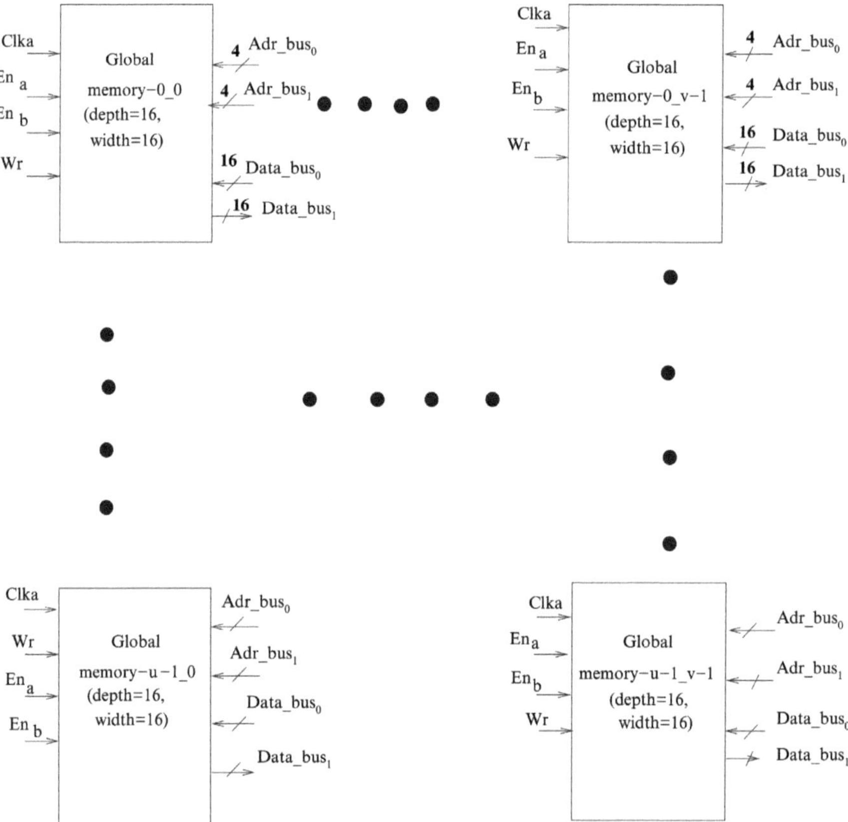

Fig. 5.10. Global memory organisation

interconnection of PEs and global memory is such that *parallel loading* of data items of different memory blocks into all PEs is possible during initialisation of place unit components.

We next describe another important element in the architecture, namely Content Addressable Memory (CAM). CAM contributes to high speed.

5.4.4 Content Addressable Memory (CAM)

We observe in Step 4 in the proposed algorithm the use of a "match-finder". This is implemented in hardware using CAMs. In our application, we have multiple CAMs (one for each x-coordinate). A sample point in the "vicinity" of a PE is found as described in Section 5.3 without operations such as squaring and square root calculation. The circuit structure is shown in Figure 5.11. Details are presented next.

For each (sampling) step along x-axis, one CAM has been assigned. The total number of CAMs corresponds to the number of steps along the x-axis. Our implementation uses 14 since we have a total of 196 sample points (at some of these points, there may

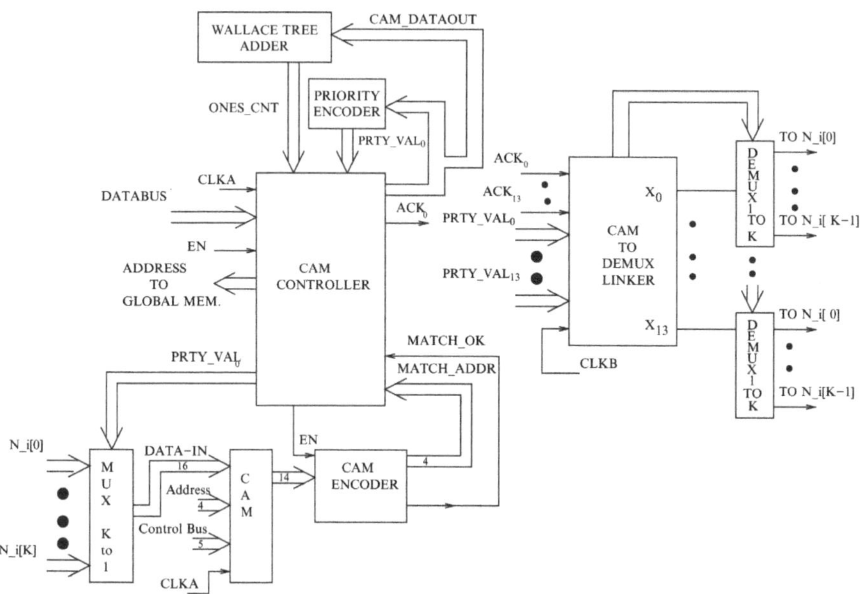

Fig. 5.11. CAM related circuits

be an obstacle preventing data collection, hence storage must be appropriately handled and is discussed below). Each CAM contains as many entries as the number of steps along the y-axis. A location in a given CAM may store the (step) value along y, if that could be explored (and no obstacle was present). Otherwise, we store zero. Each PE selects a CAM depending on the first two components of the place unit structure. If the CAM generates a MATCH_OK signal, then the PE is loaded with sensor values from the global memory. Otherwise, it is loaded with sensor values from special memory (described in subsection 5.4.5). Multiple CAM controllers, one for each CAM, are present. Each CAM controller contains a register (whose size matches the number of PEs). The appropriate bit is set by those PEs which require the address of the matched data. A Wallace tree adder is used to count number of ones in the register. A priority encoder determines for which PE, the match finding has to be carried through the controller. FPGA-based aspects of CAMs are presented in section 5.5.

5.4.5 Special Memory

While the CAM portion of the architecture presented in subsection 5.4.4 handles storage of sample points (corresponding to sampling of the environment), the special memory unit (shown in Figure 5.5) deals with storage of components for the place units (PEs). Step 5 of the proposed algorithm requires use of special memory. In the absence of obstacles, identification of a sample point in the vicinity for any place unit would be straightforward. However, when one or more obstacles exist, some of the "planned" sampling locations are unreachable - One coordinate (or both) for such a point stored in CAM would be zero. A place unit cannot be moved to such a location. Whenever a match for a sample point in the vicinity (for a place unit as per Step 4 of the algorithm) is not found, the PE (corresponding to the place unit) is loaded with the values from the special memory block from a randomly chosen address. Special memory consists of 4 blocks corresponding to sensors in the direction of $+x, -x, +y$ and $-y$.

5.4.6 Adjacency Determination Unit

In the proposed algorithm, step 6 involves connecting the appropriate place units (mapping to PEs). This is implemented as a separate unit in hardware. Figure 5.12 presents details of adjacency determination for each place unit I. $I_0, I_1, \cdots, I_5$ represent its components. A place unit S whose connectivity to I is examined has components denoted by $S_0, S_1, \cdots, S_5$. Direction information is also indicated and the internal details of the direction finder are shown in Figure 5.13. A place unit may be to the north or south of another. In general, $O, N, E, W, S, NE, NW, SE$ and SW in the figure denote respectively on (place unit on another), north, east, west, south, north-east, north-west, south-east, and south-west. The possibility of connection between a pair of place units is determined by a connection finder as shown in Figure 5.14. It must be noted that the memory requirements for Step 6 are handled by the separate set of memory blocks shown in Figure 5.5.

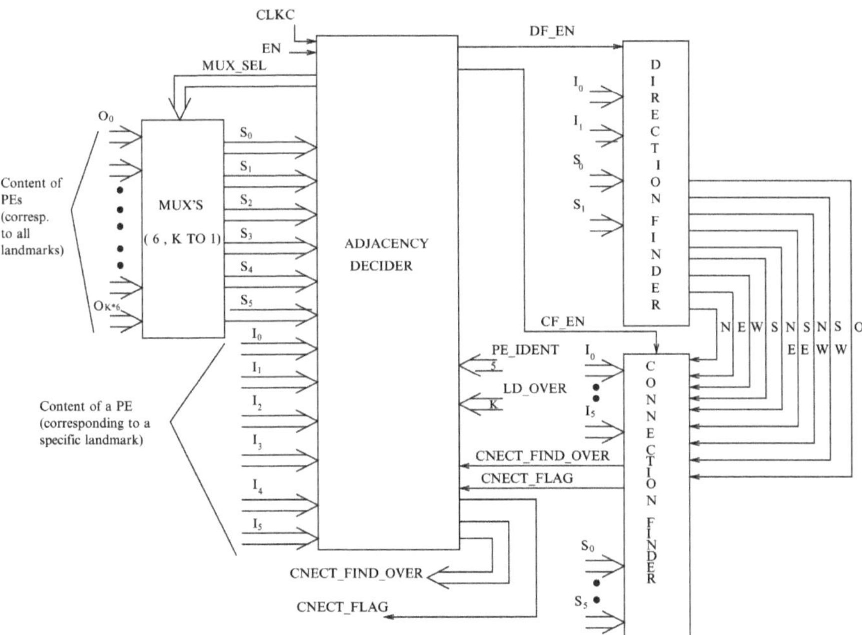

Fig. 5.12. Hardware for adjacency determination

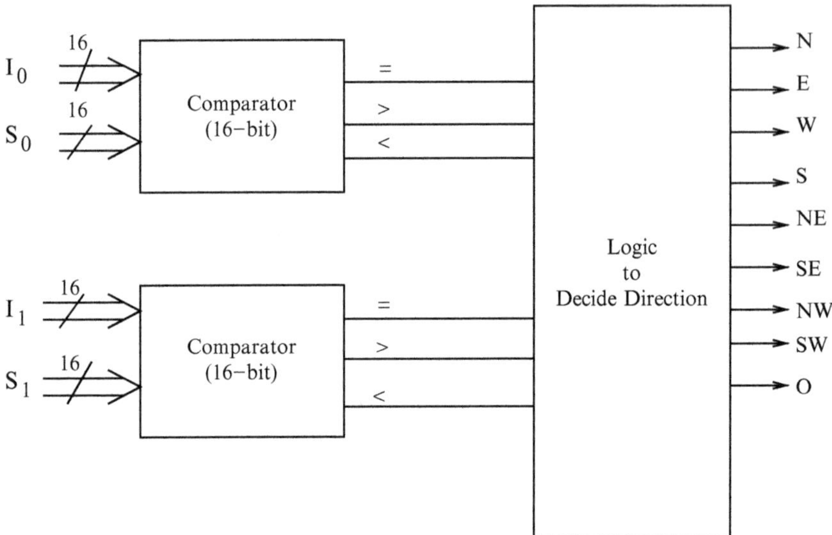

Fig. 5.13. Direction Finder in Adjacency Determination

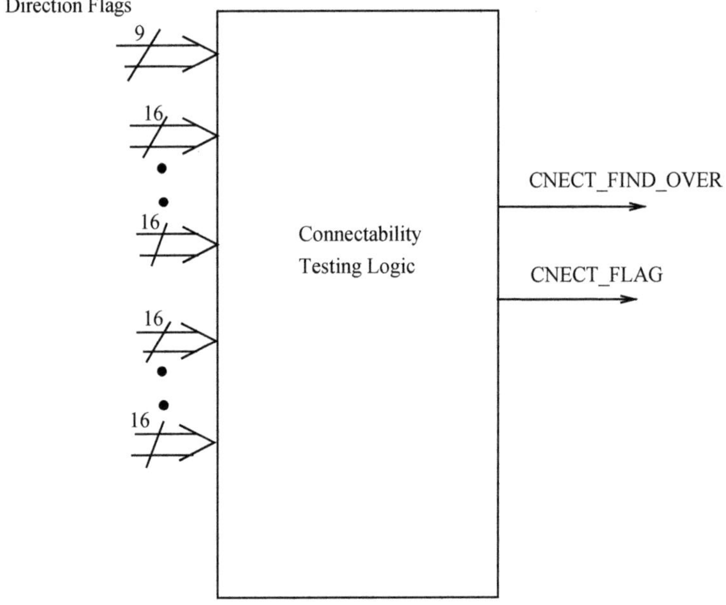

Fig. 5.14. Connection Finder in Adjacency Determination

5.4.7 Input Gating for Reducing Energy Consumption

Selective shutdown via input gating is applied to the various modules of the VLSI architecture of the Landmark Identifier and Connectivity Finder. In particular, we note the following.

- Global memory bank is active only in step 2 and step 5.
- Comparator, neighbour update and winner update are active only in step 3.
- Weight adjust controller is active only in step 4.
- Random number generator, CAM and special memory bank are active only in step 5.
- Adjacency decider and adjacency memory block are active only in step 6.

5.5 FPGA Implementation Results

VERILOG-2001 has been used for implementation. The design has been simulated using *ModelSim* and then synthesized using *Symplify 7.6*. In particular, the design has been mapped onto different devices of Xilinx to see the performance on each one of them. The specifications of the chosen devices are given in Table 5.1. General information is provided without reference to the packaging. After mapping, placement and routing of FPGA components have been carried out.

Space consumption and execution time information for the FPGA implementation for landmark computation have been gathered for the environment depicted in Figure 5.15. It is worth noting that the environment of Figure 5.15 is the same as the one for which results of exploration were presented in Chapter 4. The accessible points and the landmarks for this environment are shown in Figure 5.16.

Table 5.1. Specifications of Target FPGA Devices

Device	System Gates	CLBs	Block RAMs
XC2S200E	200K	28 x 42	14 (each 4K bit)
XCV3200E	4,074,387	104 x 156	208 (each 4K bit)
XC2V6000	6M	96 x 88	144 (each 18K bit)

Fig. 5.15. Environment on which landmark determination scheme was implemented

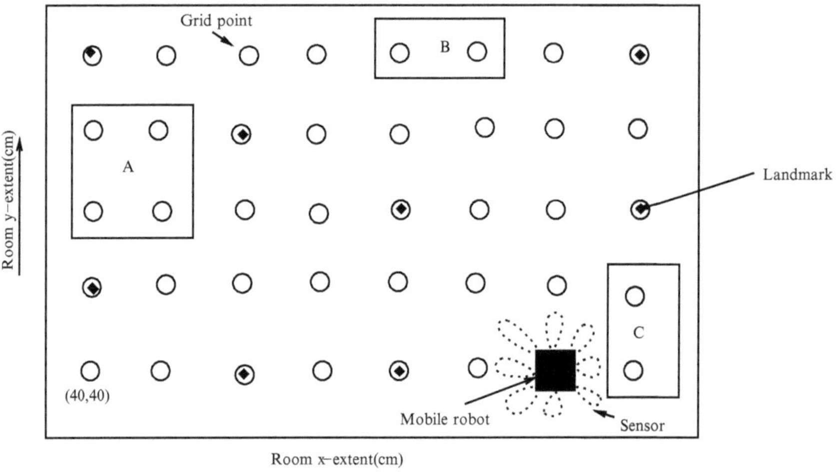

Fig. 5.16. Environment showing accessible points corresponding to Figure 5.15 and landmarks (indicated by diamonds)

Table 5.2. Results of Landmark Determination

Environment	Device	LUTs	Block RAMs
32 sample points	XC2S200E	$3,614$ (76%)	10 (71%)

Table 5.3. Results for Landmark Determination for Large Environments

Environment	Device	LUTs	Block RAMs
196 sample points	XCV3200E	$61,050$ (94%)	118 (57%)
196 sample points	XC2V6000	$60,562$ (89%)	118 (82%)

The percentage of different components of the device used by the design after placement and routing as well as the execution time are tabulated in Table 5.2. The data on execution time has been obtained from number of clock cycles required (obtained from ModelSim) and the frequency of operation. With $M = 1000, P = 32$ and $K = 8$, we have results given by Table 5.2. It is worth noting that the total space consumption on the FPGA device for landmark determination plus exploration (i.e., generation of accessible points from grid points using data from Table 4.4 in Chapter 4) is approximately 89%. This leaves approximately 10% of the XC2S200E to accommodate solution for tasks using the landmarks.

The results for a larger environment with $P = 196$ and $K = 49$ are presented in Table 5.3.

The results in Table 5.3 can be interpreted as follows. XCV3200E has a Configurable Logic Block (CLB) array of 104 x 156. It is to be noted that every CLB has 2 slices. Every slice has 2 Look-up Tables (LUTs) for a total of 64896 LUTs so the usage is 94%. Also, XCV3200E has 851,968 block RAM bits (208 4096-bit block RAMs) and the usage is 57%. Similar interpretation can be provided for usage on XC2V6000. While speedup with XC2V6000 is more, it is in general easier to fit a design into an XCV3200E than into an XC2V6000. One feature that has facilitated fitting of the entire design in one FPGA device is the following. The adjustment of the first weight vector component is done by the corresponding PE while the adjustment of the second component of all PEs is done with the help of an adjustment control unit which operates at the twice the clock frequency of the PE.

Another feature of the implementation is the use of block RAMs for handling CAMs. CAMs are built using XILINX provided primitives. The use of block RAMs for this purpose makes available the LUTs for other tasks in map construction. Global memory and special memory have also been implemented using block RAMs on Xilinx Virtex devices. This has also facilitated effective device utilization.

Information on the extent of speedup (obtained) by a custom hardware implementation has been gathered by timing a C implementation of the sequential algorithm for identification of landmarks (and adjacency information) on a general-purpose PC (1700 MHz Intel processor, on-board memory of 256 MB running Red Hat Linux 8.0). The actual running times of the sequential algorithm have been obtained via clock ticks using the *times()* function in C. The CPU time taken for finding the environment map is approximately 50 seconds when $M = 1000$, $P = 196$ and $K = 49$. For environments that are larger, the PC-based implementation typically takes close to few minutes. This amounts to a speed-up for the FPGA implementation by a factor of approximately 500. There do not appear to be any other hardware implementations for this problem in the literature and hence our comparisons are with a dedicated software solution.

Data on energy consumed by the design on Xilinx device XC2S200E is presented in Table 5.4. Energy data is valuable in the context of the life of the battery used on-board (the robot). The procedure for energy computation is as follows. First the power calculation was done using XPower tool of Xilinx. The frequency of operation (20 MHz), the Value Change Dump (vcd) file from ModelSim for the proposed design coded in Verilog and the voltage specification served as the basis for the power calculation. The

Table 5.4. FPGA Utilization and Energy Consumption Data for 32 sample points

Scenario	Freq (MHz)	LUTs	Block RAMs	Energy Consumption (micro Joules)
With BRAMs and selective shutdown	20	3614 (76%)	10 (71%)	649.6
With BRAMs and no shutdown	20	3512 (75%)	10 (71%)	1131.2

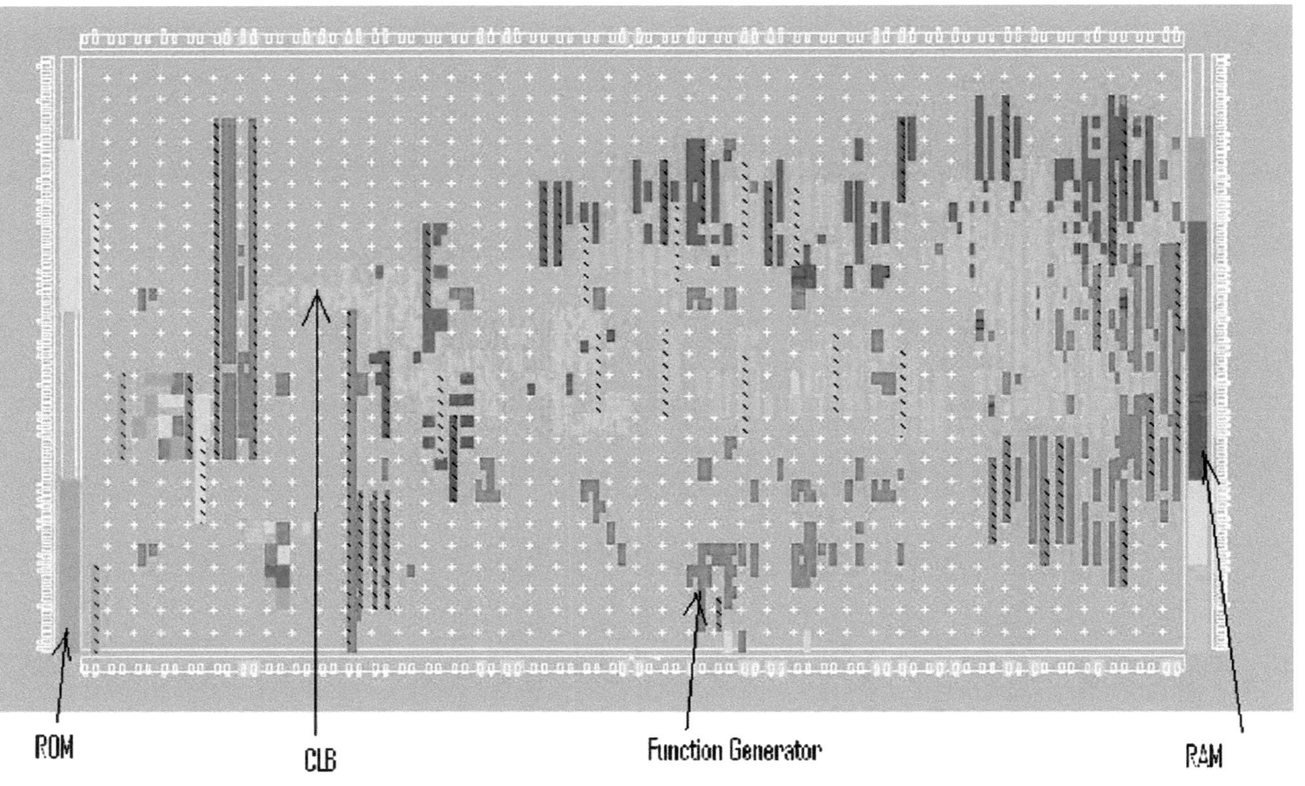

Fig. 5.17. Floorplan for implementation of landmark determination algorithm on XC2S200E

XC2S200E device consumes 333 mw power and takes 19.76 ms to complete landmark construction and adjacency determination for 32 sample points.

The floorplan for the implementation of the landmark determination scheme is shown in Figure 5.17.

5.6 Summary

This chapter has presented an efficient hardware-directed solution for landmark identification and connectivity determination in planar environments. One of the features of the approach presented is that it does *not* make any assumptions about the shape/geometry of the objects in the environment.

A new parallel algorithm with complexity $O(MP \log K)$ has been developed. Space-efficiency has also been obtained by appropriate *reuse* of hardware. The architecture developed is based on Content Addressable Memory (CAM) and Wallace-tree adders. These contribute to high performance. Further, the implementations of CAM, global memory and special memory take advantage of block RAMs present on Xilinx devices thereby making available the LUTs on the devices for other purposes. Effective device utilization has therefore been achieved.

Parallelism is particularly valuable for a dynamic environment in which replanning of paths is critical. Further, the design for an environment with 32 points fits in a low-end XC2S200E (used on the robot). For larger number of sample points (and proportionately more landmarks), the solutions fit in one XCV3200E or XC2V6000 device.

The Road Ahead

This research has studied a few aspects of robotic motion in the context of FPGA-based processing of sensor data. In particular, it has dealt with hardware-efficient exploration and landmark determination in unknown environments.

Our investigations suggest that the alternate architecture is appropriate for a wide variety of tasks and the FPGA-based approach obviates the need for additional hardware (in the form of interfaces) required in the traditional method of sensory data processing. Further, the approach indicates that for a large class of tasks, peripherals such as disks and expensive range finders (for example, laser range finders) are not required in the FPGA-based model of processing. Experiments with a prototype mobile robot equipped with an FPGA and ultrasonic sensors demonstrate the efficacy of the approach.

6.1 Contributions of this Research

The contributions of this book are as follows.

- A new *parallel* algorithm for exploration by a mobile robot in *dynamic* environments.
- A hardware-efficient architecture for exploration.
- A new *parallel* algorithm for landmark determination in planar environments.
- An efficient hardware mapping scheme for landmark determination.

K. Sridharan and P. Rajesh Kumar: *Robotic Exploration and Landmark Determination*, Studies in Computational Intelligence (SCI) **81**, 87–90 (2008)
www.springerlink.com

- A prototype of an FPGA-based mobile robot and detailed experiments on the mobile robot.

We conclude with a list of possibilities that enhance the value of the FPGA-based approach for mobile robotics.

6.2 Extensions

6.2.1 Other Types of Maps

The work presented in this book has examined hardware-efficient exploration in a grid-based environment. While this exploration process allows creation of maps using accessible grid points, alternative maps that can be constructed using sensor data are worth investigating. In particular, one geometric structure that can be constructed using merely data from ultrasonic sensors is the *Generalized Voronoi Diagram* (GVD). The GVD is the locus of points which are equidistant from two or more obstacle boundaries including the workspace boundary.

One approach to construct the GVD is using a *prediction and correction* strategy based on [27]. The approach does not involve explicit determination of position information in the form of coordinates for some reference point for the robot. It solely works on sensor readings and comparisons to make decisions at different points on the diagram as it is being constructed. A sample GVD generated using a prediction and correction strategy for a planar environment is shown in Figure 6.1.

An architecture for construction of GVD based on the prediction correction strategy would consist of a prediction circuit, correction circuit, node matching circuit, storage units for maintaining the elements of the GVD besides multiple pulse width to distance converters and command generators. In general, the usage of the FPGA device is expected to be considerably more than that for the grid-based exploration scheme presented earlier.

6.2.2 Navigation in Dynamic Environments

A variety of tasks other than exploration and landmark determination are appropriate for a mobile robot. One such task is point

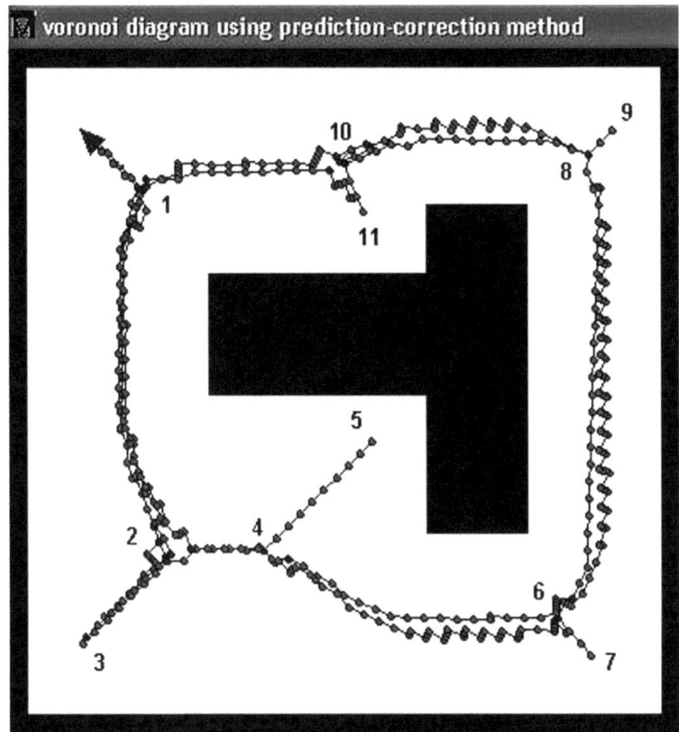

Fig. 6.1. Sensor-based GVD for an indoor environment; nodes of the GVD are labelled

to point navigation based on sensory information. Considerable work has been done on navigation in static environments. One area where architecturally efficient solutions (such as those based on FPGAs) offer promise is navigation in a *dynamic* environment. A typical scenario is illustrated in Figure 6.2.

Figure 6.2 depicts an environment with two humans and several static objects. The mobile robot moves towards point labeled G. The humans move across the path taken by the mobile robot. In particular, H_1 moves on the "left" of the robot northward while H_2 moves on the "right" also northward. The challenge for the robot is to detect the simultaneous motion of H_1 and H_2. In this context, FPGAs appear to be a promising choice in view of the ability to process data from multiple sensors *in parallel*. The scenario presented can be generalized to handle three or more objects moving in the vicinity of the robot simultaneously.

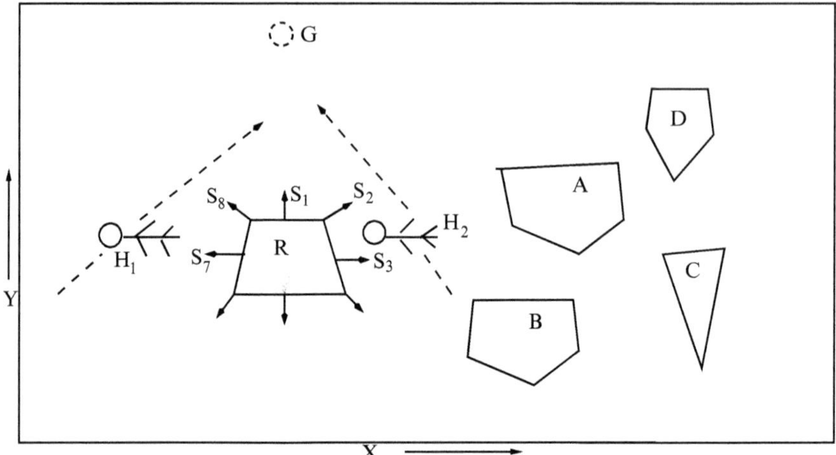

Fig. 6.2. Handling multiple moving objects during robotic navigation

6.2.3 Localization and other Tasks

The power of the FPGA-based approach can also be examined
for tasks outside the domain of exploration and navigation. In
particular, localization with limited hardware on-board a mobile
robot is one avenue for further research. Recognition of objects
keeping the sophistication (in terms of hardware) low is another
direction for further work.

6.3 Concluding Remarks

There has been an increasing emphasis on *minimalist robotics* during the last decade. Approaches to solve various problems con-
cerning mobile robots with limited sensing have been developed
[81]. The work described in this book also shares this objective.
We further believe that this work has been a step toward elegant
processing of sensor data for a large class of mobile robotic tasks.

A

Key Verilog Modules for Robotic Exploration

This appendix presents the key Verilog modules developed for implementation of the hardware-directed algorithm for exploration presented in Chapter 4. The modules have been synthesized and downloaded to the XC2S200E on the Digilab D2E board. Additional information on the Digilent board and its specifications can be obtained from www.digilentinc.com.

The top level module is presented first. The user constraint file with description of pins used on the FPGA board follows. Then the main Verilog modules are presented largely in the order in which instantiations take place in the top level module. The entire set of Verilog programs is available from `http://www.ee.iitm.ac.in/~sridhara/Verilog_ExploreAlg`.

```
//////////////////////////////////////////////////////////////////
// Module explore_env_using_dfs_128
//
// Description: Top level module for exploring indoor environment
//////////////////////////////////////////////////////////////////

module explore_env_using_dfs_128(start_exp,clk,reset1,
sen_out_cen_plus_x, sen_out_rght_plus_x,sen_out_lft_plus_x,
sen_out_plus_y,sen_out_cen_minus_x, sen_out_rght_minus_x,
sen_out_lft_minus_x,sen_out_minus_y,rs232_out,cam_on,
stack_on,co_store_on,an,ssg,ldg,cmd_gvn0);
```

```verilog
// I/O signals declaration
input start_exp; // start signal
input clk;
input reset1;

// Signals from Sensors
input sen_out_cen_plus_x; // input from sensor located at
                          //   centre of front end
input sen_out_rght_plus_x;
input sen_out_lft_plus_x;
input sen_out_plus_y;
input sen_out_cen_minus_x;
input sen_out_rght_minus_x;
input sen_out_lft_minus_x;
input sen_out_minus_y;

// Monitoring signals taken from various modules
output cmd_gvn0;
output rs232_out;
output cam_on;
output stack_on;
output co_store_on;


// I/O signals applied to FPGA  I/O Board
output [3:0] an;
output [6:0] ssg;
output ldg;

// Internal signals
wire st_cam;
wire reset;
wire [15:0] ultra_cen_plus_x;
wire [15:0] ultra_rght_plus_x;
wire [15:0] ultra_lft_plus_x;
wire [15:0] ultra_plus_y;
wire [15:0] ultra_cen_minus_x;
wire [15:0] ultra_rght_minus_x;
wire [15:0] ultra_lft_minus_x;
wire [15:0] ultra_minus_y;
wire ultra_rqst;
wire ultra_can_load;
wire [7:0] com_rs;
wire [1:0] no_of_times_tr;
wire [3:0] digit;
wire flag;
wire [15:0] outx,outy;
wire [7:0]  cnt;
```

```verilog
wire [3:0]  bit_count;
wire [6:0]  bcd0,bcd1,bcd2,bcd3;
wire [15:0] x0,y0,x1,y1;
wire [15:0] hex_x,hex_y;
wire [15:0] hex_x1,hex_y1;
wire [7:0] nxt_state;
wire [15:0] bar_x;
wire [15:0] bar_y;
wire [7:0] data;
wire [3:0] direct;
wire [3:0] visit_cnt;
wire [3:0] urt_state;
wire [3:0] send_state;
wire [7:0] data_out;
wire match_fnd;
wire stack_pop;
wire st_dly0,en_dly0,st_dly1,en_dly1;
wire uart_rst;
wire bar_rd;
wire cmd_gvn;
wire sen0_over,sen1_over,sen2_over,sen3_over,sen4_over,
     sen5_over,sen6_over,sen7_over;
wire load_dist,dis_avg;

// Initialization
assign cmd_gvn0 = ~cmd_gvn;//Command generation
                           // module monitor signal
assign reset = ~reset1;
assign ldg = 1'b1; //LED's enable signal

// Apply signals to be monitored on 7 segment displays
// to the following modules
bin_7seg bd0(bcd0,ultra_plus_y[11:8]);
bin_7seg bd1(bcd1,ultra_cen_plus_x[11:8]);
bin_7seg bd2(bcd2,nxt_state[3:0]);
bin_7seg bd3(bcd3,nxt_state[7:4]);
seven_seg sg0(ssg[0],ssg[1],ssg[2],ssg[3],ssg[4],ssg[5],ssg[6],
              an[0],an[1],an[2],an[3],clk,reset,{bcd3,bcd2,
              bcd1,bcd0});

// Instantiation of Control Unit of indoor room explorer
dyn_dfs_128 df1(start_exp,clk,reset,uart_rst,ultra_cen_plus_x,
ultra_rght_plus_x,ultra_lft_plus_x,ultra_plus_y,ultra_cen_minus_x,
ultra_rght_minus_x,ultra_lft_minus_x,ultra_minus_y,com_rs,
no_of_times_tr,cam_on,stack_on,ultra_rqst,st_dly0,en_dly0,
st_dly1,en_dly1,co_store_on,nxt_state,cmd_gvn,direct,st_cam,
sen0_over,match_fnd,stack_pop);

// Instantiation of Universal Asynchronous Transmitter Module
```

```verilog
uart ua0(clk,uart_rst,rs232_out,com_rs,no_of_times_tr,cmd_gvn,
data,urt_state,send_state,data_out,stack_pop);

// Instantiation of Pulse Width to Distance converters
ultra_sonic_sensor ul0(clk,ultra_rqst,sen_out_cen_plus_x,
ultra_cen_plus_x,sen0_over);
ultra_sonic_sensor u9(clk,ultra_rqst,sen_out_rght_plus_x,
ultra_rght_plus_x,sen1_over);
ultra_sonic_sensor u8(clk,ultra_rqst,sen_out_lft_plus_x,
ultra_lft_plus_x,sen2_over);
ultra_sonic_sensor ul1(clk,ultra_rqst,sen_out_plus_y,
ultra_plus_y,sen3_over);
ultra_sonic_sensor ul2(clk,ultra_rqst,sen_out_cen_minus_x,
ultra_cen_minus_x,sen4_over);
ultra_sonic_sensor ul3(clk,ultra_rqst,sen_out_rght_minus_x,
ultra_rght_minus_x,sen5_over);
ultra_sonic_sensor ul4(clk,ultra_rqst,sen_out_lft_minus_x,
ultra_lft_minus_x,sen6_over);
ultra_sonic_sensor ul5(clk,ultra_rqst,sen_out_minus_y,
ultra_minus_y,sen7_over);

// Instantiation of different delay modules which are used
// to delay the distance measurements for distinguishing
// dynamic obstacles from static obstacles
delay d0(st_dly0,en_dly0,clk);
more_delay d1(st_dly1,en_dly1,clk);

endmodule


# UCF for mapping to FPGA on Digilent board
NET "start_exp"            LOC =   "P16";
NET "clk"                  LOC =   "P80";
NET "reset1"               LOC =   "p40";
NET "sen_out_cen_plus_x"   LOC =   "p175";
NET "sen_out_rght_plus_x"  LOC =   "p138";
NET "sen_out_lft_plus_x"   LOC =   "p166";
NET "sen_out_minus_y"      LOC =   "p132";
NET "sen_out_cen_minus_x"  LOC =   "p146";
NET "sen_out_rght_minus_x" LOC =   "p127";
NET "sen_out_lft_minus_x"  LOC =   "p150";
NET "sen_out_plus_y"       LOC =   "p152";
NET "rs232_out"            LOC =   "P201";
NET "stack_on"          LOC = "p44";#led0
NET "cam_on"            LOC = "p46";#led1
NET "co_store_on"       LOC = "p48";#led2
NET "cmd_gvn0"          LOC = "P57"; #led3
NET "ldg"           LOC = "P68";
NET "ssg<0>"               LOC =   "P17";
```

```
NET "ssg<1>"                    LOC =   "P20";
NET "ssg<2>"                    LOC =   "P22";
NET "ssg<3>"                    LOC =   "P24";
NET "ssg<4>"                    LOC =   "P29";
NET "ssg<5>"                    LOC =   "P31";
NET "ssg<6>"                    LOC =   "P34";
NET "an<0>"                     LOC =   "P45";
NET "an<1>"                     LOC =   "P47";
NET "an<2>"                     LOC =   "P49";
NET "an<3>"                     LOC =   "P56";


/////////////////////////////////////////////////////////
// Module dyn_dfs_128
//
// Description: Control unit for exploration
///////////////////////////////////////////////////////////////

module dyn_dfs_128(start_exp,clk,rst,uart_rst,ultra_cen_plus_x,
ultra_rght_plus_x,ultra_lft_plus_x,ultra_plus_y,
ultra_cen_minus_x,ultra_rght_minus_x,ultra_lft_minus_x,
ultra_minus_y,com_rs,no_of_times_tr,cam_on,stack_on,ultra_rqst,
st_dly0,en_dly0,st_dly1,en_dly1,co_store_on,nxt_state,cmd_gvn,
direct,st_cam,sen0_over,match_fnd,stack_pop);

// I/O signals declaration
input start_exp;// Signal for starting exploration
input clk;
input rst;

// Signals coming from ultrasonic range finders
input [15:0] ultra_cen_plus_x,ultra_rght_plus_x;
input [15:0] ultra_lft_plus_x,ultra_plus_y;
input [15:0] ultra_cen_minus_x,ultra_rght_minus_x;
input [15:0] ultra_lft_minus_x,ultra_minus_y;

// Signals monitored
output stack_on;
output cam_on;
output    [7:0] com_rs;
output [1:0] no_of_times_tr;
output reg ultra_rqst;
output reg st_dly0;
input en_dly0;
output reg st_dly1;
input en_dly1;
output reg uart_rst;
output co_store_on;
output [7:0] nxt_state;
input cmd_gvn;
```

```verilog
output [3:0] direct;
output st_cam;
input sen0_over;
output match_fnd;
output reg stack_pop;

// Signals for memory modules
reg [5:0] visit_cnt;

// Signals used for shutting down of adjacency memory blocks
reg adjn_en0;
reg adjn_en1;
reg adjn_en2;
reg adjn_en3;

// Signals used for writing into adjacency memory blocks
reg adjn_wr0;
reg adjn_wr1;
reg adjn_wr2;
reg adjn_wr3;

// Address buses of adjacency memory blocks
reg [7:0] adjn_add_bus0;
reg [7:0] adjn_add_bus1;
reg [7:0] adjn_add_bus2;
reg [7:0] adjn_add_bus3;

// Databuses of adjacency memory blocks
wire [7:0] adjn_out_data_bus0;
wire [7:0] adjn_out_data_bus1;
wire [7:0] adjn_out_data_bus2;
wire [7:0] adjn_out_data_bus3;

// Indicators(ok/otherwise) of adjacency memory blocks
wire adjn_on0;
wire adjn_on1;
wire adjn_on2;
wire adjn_on3;

// Data bus of dyn_dfs_128 module
reg [7:0] data_in;

// Signals of stack
reg s_push;
wire s_empty;
reg s_read;
reg s_pop;
wire [7:0] s_data_in;
wire [7:0] s_data_out;
```

```verilog
wire [7:0]stack_add;
wire mem_on;

// States declaration
parameter st0=0,st1=1,st2=2,st3=3,st4=4,st5=5,st6=6,st7=7,
          st8=8,st9=9,st10=10,st11=11,st12=12,st13=13,
          st14=14,st15=15,st16=16,st17=17,st18=18,st19=19,
          st20=20,st21=21,st22=22,st23=23,st24=24,st25=25,
          st26=26,st27=27,st28=28,st29=29,st30=30,st31=31,
          st32=32,st33=33,st34=34,st35=35,st36=36,st37=37,
          st38=38,st39=39,st40=40;

// Internal registers of dyn_dfs-128
reg [5:0] state;
reg [3:0] adj_finder;
reg[7:0] cur_node_pntr;
reg[7:0] nxt_node_pntr;
reg[3:0] adj_test;
reg [2:0] adj_nd_cnt;//adjacent node count
reg mov_plus_x;
reg mov_plus_y;
reg mov_minus_x;
reg mov_minus_y;
reg [7:0] cam_addr;
reg [15:0] x_crd;
reg [15:0] y_crd;
reg [6:0] store_addr;

// Signal used for shutting down of CAM
reg cam_en;

// Signals  of CAM read/write
reg cam_wr_en;
reg cam_er_wr;
reg wr_ram;
reg match_en;
reg match_rst;
wire  match;
reg [2:0] int_cnt;

// Signal used for shutting down of visted_node_store_128
// memory block
reg vis_node_en;

// Signals used for read/write operations of
// visted_node_store_128 memory block
reg [6:0] vis_node_add_bus;
reg [7:0] vis_node_in_data_bus;
wire [7:0] vis_node_out_data_bus;
```

```verilog
wire vis_on;
wire xcrd_on;
wire ycrd_on;
reg vis_wr;

// Signal used for shutting down of
// node_x_cord_store_128 memory block
reg node_x_crd_en;

// Signals used for read/write operations of
// node_x_cord_store_128 memory block
reg node_x_wr;
reg [6:0] node_x_crd_add_bus;
reg [15:0] node_x_crd_in_data_bus;
wire [15:0] node_x_crd_out_data_bus;

// Signal used for shutting down of
// node_y_cord_store_128 memory block
reg node_y_crd_en;

// Signals used for read/write operations of
// node_y_cord_store_128 memory block
reg node_y_wr;
reg [6:0] node_y_crd_add_bus;
reg [15:0] node_y_crd_in_data_bus;
wire [15:0] node_y_crd_out_data_bus;
reg [1:0] cur_direct;

// Signal used for shutting down of
// nxt_node_dir_store_128 memory block
reg node_dir_en;

// Signals used for read/write operations of
// nxt_node_dir_store_128 memory block
reg node_dir_wr;
reg [7:0] node_dir_add_bus;
reg [1:0] node_dir_in_data_bus;
wire [1:0] node_dir_out_data_bus;
wire dir_store_on;
reg poped_node;

// Signal used for shutting down of
// sensor_plusx_cord_store_128 memory block
reg plus_x_crd_en;

// Signals for read/write operations of
// sensor_plusx_cord_store_128 memory block
reg plus_x_wr;
reg [6:0] plus_x_crd_add_bus;
```

```
reg [15:0] plus_x_crd_in_data_bus;
wire [15:0] plus_x_crd_out_data_bus;
wire plus_x_crd_on;

// Signal used for shutting down of
// sensor_plusy_cord_store_128 memory block
reg plus_y_crd_en;

// Signals for read/write operations of
// sensor_plusy_cord_store_128 memory block
reg plus_y_wr;
reg [6:0] plus_y_crd_add_bus;
reg [15:0] plus_y_crd_in_data_bus;
wire [15:0] plus_y_crd_out_data_bus;
wire plus_y_crd_on;

// Signal used for shutting down of
// sensor_minusx_cord_store_128 memory block
reg minus_x_crd_en;

// Signals for read/write operations of
// sensor_minusx_cord_store_128 memory block
reg minus_x_wr;
reg [6:0] minus_x_crd_add_bus;
reg [15:0] minus_x_crd_in_data_bus;
wire [15:0] minus_x_crd_out_data_bus;
wire minus_x_crd_on;

// Signal used for shutting down of
// sensor_minusy_cord_store_128 memory block
reg minus_y_crd_en;

// Signals for read/write operations of
// sensor_minusy_cord_store_128 memory block
reg minus_y_wr;
reg [6:0] minus_y_crd_add_bus;
reg [15:0] minus_y_crd_in_data_bus;
wire [15:0] minus_y_crd_out_data_bus;
wire minus_y_crd_on;

// Signal  for shutting down of cmd_gen_128 block
// (cmd generator for stepper motor driver)
reg cmd_en;

// Signal for controlling cmd_gen_128 block
// (command generator for stepper motor driver)
wire upx,upy,umx,umy;
reg match_fnd;
```

```verilog
// Instantiation of CAM to store index of nodes visited
CAM_RAMB_128 ca0(cam_en,data_in,cam_addr,cam_wr_en,cam_er_wr,
wr_ram,clk,match_en,match_rst,match);

// Instantiation of memory blk to store adj structure of
// environment explored
adja_nodes_store0_128 ans0(clk,adjn_en0,adjn_add_bus0,
adjn_out_data_bus0,adjn_on0) ;
adja_nodes_store1_128 ans1(clk,adjn_en1,adjn_add_bus1,
adjn_out_data_bus1,adjn_on1) ;
adja_nodes_store2_128 ans2(clk,adjn_en2,adjn_add_bus2,
adjn_out_data_bus2,adjn_on2) ;
adja_nodes_store3_128 ans3(clk,adjn_en3,adjn_add_bus3,
adjn_out_data_bus3,adjn_on3) ;

// Instantiation of mem blks to store index of visited
// nodes and their coordinates
visted_node_store_128 vis(clk,vis_node_en,vis_wr,
vis_node_add_bus,vis_node_in_data_bus,
vis_node_out_data_bus,vis_on) ;
node_x_cord_store_128 xcrd(clk,node_x_crd_en,node_x_wr,
node_x_crd_add_bus,node_x_crd_in_data_bus,
node_x_crd_out_data_bus,xcrd_on) ;
node_y_cord_store_128 ycrd(clk,node_y_crd_en,node_y_wr,
node_y_crd_add_bus,node_y_crd_in_data_bus,
node_y_crd_out_data_bus,ycrd_on) ;

// Instantiation of stack for storing visited nodes
// which will be used during back tracking
stack_128 stk0(clk,rst,data_in,s_push,s_empty,s_read,s_pop,
s_data_in,s_data_out,mem_on,stack_add);

// Instantiation of command generator to generate
// required commands for moving mobile robot to next node
cmd_gen_128 cmd0(clk,rst,mov_plus_x,mov_plus_y,
mov_minus_x,mov_minus_y,com_rs,no_of_times_tr,direct);


// These assignments for testing the desired signals
assign adj_nodes_store_on = adjn_on0 | adjn_on1
| adjn_on2 | adjn_on3;
assign stack_on = s_empty;
assign co_store_on = vis_on | xcrd_on | ycrd_on
| dir_store_on;
assign nxt_state = {2'b00,state};
assign cam_on =  adj_nodes_store_on | plus_x_crd_on | plus_y_crd_on
| minus_x_crd_on | minus_y_crd_on;
assign st_cam = match;
```

```verilog
always @(posedge clk or negedge rst)
begin
if(!rst)// Initialisation of signals and internal regs
begin

// Enabling adjacency memories
adjn_en0 = 1'b0;
adjn_en1 = 1'b0;
adjn_en2 = 1'b0;
adjn_en3 = 1'b0;

// Initializing register storing  adjacency information
adj_finder = 4'b0;

// Initializing stack
s_push = 1'b0;
s_pop = 1'b0;
s_read = 1'b0;
store_addr = 7'b0;

// Initializing pointers keeping node IDs
cur_node_pntr = 8'b0;
nxt_node_pntr = 8'b0;

// Initializing direction of robot
mov_plus_x = 1'b0;
mov_plus_y= 1'b0;
mov_minus_x = 1'b0;
mov_minus_y = 1'b0;

// Initializing CAM control signals
cam_en = 1'b1;
cam_wr_en = 1'b0;
cam_er_wr = 1'b0;
wr_ram = 1'b0;
match_en = 1'b0;
match_rst = 1'b0;
cam_addr = 8'b0;

// Initialization of ultrasonic sensor,UART and delays
ultra_rqst = 1'b0;
uart_rst = 1'b1;
st_dly0 = 1'b0;
st_dly1 = 1'b0;


// Initialization of direction store memory
node_dir_en = 1'b1;
node_dir_wr = 1'b0;
```

```verilog
node_dir_add_bus = 8'b0;

// Initialization of coordinates
x_crd = 16'b0;
y_crd = 16'b0;
node_x_crd_en = 1'b1;
node_y_crd_en = 1'b1;
node_x_wr = 1'b0;
node_y_wr = 1'b0;
node_x_crd_add_bus = 8'b0;
node_y_crd_add_bus = 8'b0;
poped_node = 1'b0;
vis_node_en = 1'b1;
vis_node_add_bus = 7'b0;

// Sensor value storage
minus_x_crd_en = 1'b1;
minus_x_wr = 1'b1;
minus_x_crd_add_bus = 16'b0;
minus_y_crd_en = 1'b1;
minus_y_wr = 1'b1;
minus_y_crd_add_bus = 16'b0;
plus_x_crd_en = 1'b1;
plus_x_wr = 1'b1;
plus_x_crd_add_bus = 16'b0;
plus_y_crd_en = 1'b1;
plus_y_wr = 1'b1;
plus_y_crd_add_bus = 16'b0;

// Signal for enable for Command generator
cmd_en = 1'b0;

// Signal for indicating  stack pop
stack_pop = 1'b0;

// Enable coordinates storage
state = st0;

end

else
begin
case(state)

// Wait state for receiving ultrasonic sensors outputs
// Initialisation of pulse width to distance
// converter,CAM and stack
st0: begin
    cam_en = 1'b0;
```

```verilog
        match_en = 1'b1;
        match_rst = 1'b0;
        cam_wr_en = 1'b0;
        cam_er_wr = 1'b0;
        wr_ram = 1'b0;
        ultra_rqst = 1'b0;
        match_fnd = 1'b0;
        s_push = 1'b0;
        s_pop = 1'b0;
        s_read = 1'b0;

        if(start_exp)
        state = st1;
        else
        state = st0;
        end

// Storing x and y co-ordinates of grid point
st1: begin
        state = st2;
        cam_en = 1'b1;
        st_dly0 = 1'b1;
        vis_wr = 1'b1;
        node_x_wr = 1'b1;
        node_y_wr = 1'b1;
        node_x_crd_in_data_bus = x_crd;
        node_y_crd_in_data_bus = y_crd;
        end

// Storing grid point index in CAM and STACK
st2: begin
        mov_minus_x = 1'b0;
        mov_minus_y = 1'b0;
        mov_plus_x = 1'b0;
        mov_plus_y = 1'b0;
        adj_finder = 4'b0000;
        data_in = cur_node_pntr;
        vis_node_in_data_bus = cur_node_pntr;
        vis_wr = 1'b0;
        node_x_wr = 1'b0;
        node_y_wr = 1'b0;
        st_dly0 = 1'b0;
        s_push = 1'b1;
        cam_wr_en = 1'b1;
        cam_er_wr = 1'b1;
        wr_ram = 1'b1;
        match_en = 1'b0;
        match_rst = 1'b1;
        state = st3;
```

```verilog
          end

  st3: begin
       s_push = 1'b0;
       cam_wr_en = 1'b1;
       cam_er_wr = 1'b1;
       wr_ram = 1'b0;
       state = st4;
       end

  // Moving pointer to next location of CAM
  st4: begin
       cam_wr_en = 1'b0;
       cam_er_wr = 1'b0;
       wr_ram = 1'b0;
       cam_addr = cam_addr + 1'b1;
       state = st5;
       end

  // Measuring distances to the nearest obstacles
  // in 8 directions
  st5: begin
       ultra_rqst = 1'b1;
       if(sen0_over)
       begin
       if(ultra_cen_plus_x > 16'h150)
       begin
       adj_finder = adj_finder | 4'b1000;
       state = st6;
       end
       else
       begin
       adj_finder = adj_finder & 4'b0111;
       state = st7;
       end
       end
       else
       state = st5;
       end

  st6: begin
       if(en_dly0)
       begin
       state = st7;
       end
       else
       begin
       ultra_rqst = 1'b1;
       state = st6;
```

```verilog
        end
    end

// Adjacency finding and storing node in CAM and stack
// Determining the direction in which robot can move on
// the basis of priority

st7: begin
    ultra_rqst = 1'b0;
    adjn_en0 = 1'b1;
    adjn_en1 = 1'b1;
    adjn_en2 = 1'b1;
    adjn_en3 = 1'b1;
    node_dir_en = 1'b1;
    node_x_crd_en = 1'b1;
    node_y_crd_en = 1'b1;
    vis_node_en = 1'b1;
    minus_x_crd_en = 1'b1;
    minus_y_crd_en = 1'b1;
    plus_x_crd_en = 1'b1;
    plus_y_crd_en = 1'b1;
    if(visit_cnt<6'b110110)
    adj_finder = adj_finder | 4'b0001;
    if(ultra_cen_plus_x > 16'h200)
    adj_finder = adj_finder | 4'b1000;
    if(ultra_plus_y >= 16'h300)
    adj_finder = adj_finder | 4'b0100;
    if(ultra_cen_minus_x > 16'h300)
    adj_finder = adj_finder | 4'b0010;
    state = st8;
    end

// If it is not possible to visit any direction then
// test stack for backtracking or for stopping the robot

st8: begin
    store_addr = store_addr + 1'b1;
    if(s_empty)
    state = st39;
    else
    state = st9;
    end

st9: begin
    adj_test=adj_finder;
    adj_nd_cnt = adj_finder[0] + adj_finder[1]
                 + adj_finder[2] + adj_finder[3];
    state = st10;
    end
```

```verilog
st10: begin
      if(adj_nd_cnt != 3'b0)
      begin
      mov_minus_x = 1'b0;
      mov_minus_y = 1'b0;
      mov_plus_x = 1'b0;
      mov_plus_y = 1'b0;
      node_dir_wr = 1'b1;
      state = st11;
      end
      else
      begin
      mov_minus_x = 1'b0;
      mov_minus_y = 1'b0;
      mov_plus_x = 1'b0;
      mov_plus_y = 1'b0;
      state = st18;
      end
      end

// Decide the direction in which the robot can move
st11: begin
      if(adj_test[3] == 1'b1)
      begin
      mov_plus_x = 1'b1;
      node_dir_in_data_bus = 2'd0;
      adjn_en0  = 1'b1;
      adjn_add_bus0 = cur_node_pntr;
      end
      else if(adj_test[1] == 1'b1)
      begin
      node_dir_in_data_bus = 2'd2;
      mov_minus_x = 1'b1;
      adjn_en2= 1'b1;
      adjn_add_bus2 = cur_node_pntr;
      end
      else if(adj_test[0] == 1'b1)
      begin
      node_dir_in_data_bus = 2'd3;
      mov_minus_y = 1'b1;
      adjn_en3= 1'b1;
      adjn_add_bus3 = cur_node_pntr;
      end
      else if(adj_test[2] == 1'b1)
      begin
      mov_plus_y = 1'b1;
      node_dir_in_data_bus = 2'd1;
      adjn_en1= 1'b1;
```

```
        adjn_add_bus1 = cur_node_pntr;
        end
        node_dir_wr = 1'b0;
        state = st12;
        end

st12: begin
        state = st13;
        end

st13: begin
        if(adj_test[3] == 1'b1)
        nxt_node_pntr = adjn_out_data_bus0;
        else if(adj_test[1] == 1'b1)
        nxt_node_pntr = adjn_out_data_bus2;
        else if(adj_test[0] == 1'b1)
        nxt_node_pntr = adjn_out_data_bus3;
        else if(adj_test[2] == 1'b1)
        nxt_node_pntr = adjn_out_data_bus1;
        state = st14;
        end

st14: begin
        data_in = nxt_node_pntr;
        match_en = 1'b1;
        match_rst = 1'b1;
        cam_wr_en = 1'b0;
        cam_er_wr = 1'b0;
        wr_ram = 1'b0;
        state = st15;
        end

st15: state = st16;

st16: begin
        state = st17;
        end

// Verify whether the node in the decided direction of
// visit has been already visited
st17: begin
        if(match)
        begin
        match_fnd = 1'b1;
        if(mov_plus_x)
        adj_test[3] = 1'b0;
        else if(mov_plus_y)
        adj_test[2] = 1'b0;
        else if(mov_minus_x)
```

```verilog
                adj_test[1] = 1'b0;
                else if(mov_minus_y)
                adj_test[0] = 1'b0;
                adj_nd_cnt = adj_nd_cnt - 1'b1;
                state =st10;
                end
                else
                begin
                match_fnd = 1'b0;
                node_dir_add_bus = node_dir_add_bus + 1'b1;
                state = st32;
                end
                end

        // Popping from stack for backtrack
        st18: begin
                node_dir_add_bus = node_dir_add_bus - 1'b1;
                s_pop = 1'b1;
                poped_node = 1'b1;
                state = st19;
                end

        st19: begin
                s_pop = 1'b0;
                state = st20;
                end

        st20: begin
                state =st21;
                end

        st21: state = st22;

        st22: begin
                s_pop = 1'b1;
                poped_node = 1'b1;
                state = st23;
                end

        st23: begin
                s_pop = 1'b0;
                state = st24;
                end

        st24: begin
                state =st25;
                end

        st25: state = st26;
```

```verilog
// Testing stack: if it is empty, stop the robot

st26: begin
      if(s_empty)
      state = st39;
      else
      state = st27;
      end

st27: begin
      s_read = 1'b1;
      state = st28;
      end

st28: begin
      s_read = 1'b0;
      state = st29;
      end

st29: begin
      mov_minus_x = 1'b0;
      mov_minus_y = 1'b0;
      mov_plus_x = 1'b0;
      mov_plus_y = 1'b0;
      state = st30;
      end

st30: state = st31;

st31: begin
      nxt_node_pntr = s_data_out;
      data_in = s_data_out;
      if(node_dir_out_data_bus == 2'd0)
      mov_minus_x = 1'b1;
      else if( node_dir_out_data_bus== 2'd1)
      mov_minus_y = 1'b1;
      else if(node_dir_out_data_bus == 2'd2)
      mov_plus_x = 1'b1;
      else if(node_dir_out_data_bus == 2'd3)
      mov_plus_y = 1'b1;
      state = st32;
      end

st32: begin
      cmd_en = 1'b1;
      cur_node_pntr = nxt_node_pntr;
      data_in = nxt_node_pntr;
      uart_rst = 1'b0;
```

```
      state = st33;
      end


st33: begin
      if(cmd_gvn)
      begin
      uart_rst = 1'b1;
      cmd_en = 1'b0;
      state = st34;
      end
      else
      state = st33;
      end


st34: begin
      state = st35;
      end


st35: begin
      visit_cnt = visit_cnt + 1'b1;
      if(poped_node)
      begin
      cam_en = 1'b0;
      poped_node = 1'b0;
      data_in = cur_node_pntr;
      stack_pop = 1'b1;
      state = st1;
      end
      else
      begin
      node_x_crd_add_bus = node_x_crd_add_bus + 1'b1;
      node_y_crd_add_bus = node_y_crd_add_bus + 1'b1;
      plus_x_crd_add_bus = plus_x_crd_add_bus + 1'b1;
      plus_y_crd_add_bus = plus_y_crd_add_bus + 1'b1;
      minus_x_crd_add_bus = minus_x_crd_add_bus + 1'b1;
      minus_y_crd_add_bus = minus_y_crd_add_bus + 1'b1;
      vis_node_add_bus = vis_node_add_bus + 1'b1;
      state = st36;
      end
      end


st36: begin
      st_dly1=1'b1;
      state = st37;
      end


st37: begin

// Initialization
```

```
adjn_en0 = 1'b0;
adjn_en1 = 1'b0;

adjn_en2 = 1'b0;
adjn_en3 = 1'b0;
cam_en = 1'b0;
node_x_crd_en = 1'b0;
node_y_crd_en = 1'b0;
vis_node_en = 1'b0;
minus_x_crd_en = 1'b0;
minus_y_crd_en = 1'b0;
plus_x_crd_en = 1'b0;
plus_y_crd_en = 1'b0;
st_dly1=1'b0;
state = st38;
end

st38: begin
      if(en_dly1)
      begin
      ultra_rqst = 1'b1;
      state=st1;
      end
      else
      state=st38;
      end

st39: state = st39;

endcase

end

end

endmodule

//////////////////////////////////////////////////////
//  Module cmd_gen_128
//
// Description: Command generation module for stepper motors
//////////////////////////////////////////////////////////////

module cmd_gen_128(clk,rst,mov_plus_x,mov_plus_y,mov_minus_x,
mov_minus_y,com_rs,no_of_times_tr,direct);

// I/O declarations
```

```verilog
input clk;
input rst;

//Flags for moving the mobile robot in any one of
//  +X, +Y, -X and -Y directions
input mov_plus_x;
input mov_plus_y;
input mov_minus_x;
input mov_minus_y;

// Control signals for stepper motor driver
output reg[1:0] no_of_times_tr; // number of times of
                                // rotation for motor
output reg [7:0]com_rs;// declaration of cmd word for UAT
output [3:0] direct;

// Initialization of direct signal for turning the robot
assign direct = {mov_plus_x,mov_plus_y,mov_minus_x,mov_minus_y};

// Generation of control signals as per the
// value of direction
always @(posedge clk or negedge rst)
begin
if(!rst)
begin
com_rs = 8'd0;
no_of_times_tr = 2'd0;
end
else
begin
case(direct)

// commands for moving robot forward/backward
// and rotating CW/CCW
4'd1: begin
    com_rs = 8'h72;
    no_of_times_tr = 2'd1;
    end

4'd2: begin
    com_rs = 8'h66;
    no_of_times_tr = 2'd1;
    end

4'd4: begin
    com_rs = 8'h6c;
    no_of_times_tr = 2'd1;
    end
```

```verilog
4'd8: begin
      com_rs = 8'h62;
      no_of_times_tr = 2'd0;
      end

endcase
end
end

endmodule


////////////////////////////////////////////////////////
// Module node_x_cord_store_128
// Description: BLOCK RAM to store x-coordinates of nodes
////////////////////////////////////////////////////////

module node_x_cord_store_128 (clk,node_x_crd_en,node_x_wr,
node_x_crd_add_bus,node_x_crd_in_data_bus,
node_x_crd_out_data_bus,xcrd_on) ;

// I/O declarations
input clk;
input  node_x_crd_en;// Signal for enabling memory
input node_x_wr;
input [6:0] node_x_crd_add_bus;
input [15:0] node_x_crd_in_data_bus;
output [15:0] node_x_crd_out_data_bus;
output reg xcrd_on;//Signal for monitoring
reg [15 :0] ram [127:0];
reg [6:0] read_a;

always @(posedge clk)
begin
if (node_x_crd_en)
begin
xcrd_on  = 1'b1;
if (node_x_wr)
ram[node_x_crd_add_bus] <= node_x_crd_in_data_bus;
read_a <= node_x_crd_add_bus;
end
else
xcrd_on  = 1'b0;
end

assign node_x_crd_out_data_bus = ram[read_a];

endmodule
```

```verilog
///////////////////////////////////////////////////////////
// Module node_y_cord_store_128
//  Description:  BLOCK RAM to store y-coordinate of nodes
///////////////////////////////////////////////////////////

module node_y_cord_store_128 (clk,node_y_crd_en,node_y_wr,
node_y_crd_add_bus,node_y_crd_in_data_bus,
node_y_crd_out_data_bus,ycrd_on) ;

// I/O declarations
input clk;
input  node_y_crd_en; // Signal for enable
input node_y_wr;
input [6:0] node_y_crd_add_bus;
input [15:0] node_y_crd_in_data_bus;
output [15:0] node_y_crd_out_data_bus;
output reg ycrd_on; // Signal for monitoring
reg [15 :0] ram [127:0];
reg [6:0] read_a;

always @(posedge clk)
begin
if (node_y_crd_en)
begin
ycrd_on  <= 1'b1;
if (node_y_wr)
ram[node_y_crd_add_bus] <= node_y_crd_in_data_bus;
read_a <= node_y_crd_add_bus;
end
else
ycrd_on  <= 1'b0;
end

assign node_y_crd_out_data_bus = ram[read_a];

endmodule


///////////////////////////////////////////////////////////
// Module stack_128
//
// Description: Stack control unit
///////////////////////////////////////////////////////////

'define s_width 8  // Width of stack

module stack_128(clk,rst,data_in,s_push,s_empty,
s_read,s_pop,s_data_in,s_data_out,mem_on,stack_add);
```

```verilog
// I/O declarations
input clk;
input rst;
input [('s_width - 1) :0] data_in;
input  s_push;
output  s_empty;
input s_read;
input s_pop;
output  reg [('s_width - 1) :0] s_data_in;
output  [('s_width - 1) :0] s_data_out;
output [7:0] stack_add;
output mem_on;//Signal for monitoring stack memory

parameter st0=0,st1=1,st2=2,st3=3,st4=4,st5=5,st6=6,st7=7;

reg [2:0] state;
reg intrnl_en;
reg intrnl_wr;
reg [6:0]  intrnl_add_bus;
wire mem_en,mem_wr;
wire [6:0]  mem_add_bus;
wire [('s_width - 1) :0] mem_in_data_bus;
wire [('s_width - 1) :0] mem_out_data_bus;

assign mem_en = intrnl_en;
assign mem_wr = intrnl_wr;
assign mem_add_bus = intrnl_add_bus;
assign mem_in_data_bus = s_data_in;
assign s_empty = (intrnl_add_bus == 7'b1111111) ? 1'b1 : 1'b0;
assign stack_add = {1'b0,intrnl_add_bus};
assign s_data_out = mem_out_data_bus;

stack_memblk_128 stm0(clk,mem_en,mem_wr,mem_add_bus,
mem_in_data_bus,mem_out_data_bus,mem_on);

always @(posedge clk or negedge rst)
begin
if(!rst) //Initialization of stack top to the
        // highest available address
begin
intrnl_add_bus = 7'b1111111;
intrnl_wr = 1'b0;
intrnl_en = 1'b0;
state = st0;
end
else
begin
case(state)
```

```verilog
st0: begin // Testing level of input signals and
          // then taking actions
     if(s_push)
     state = st1;
     else if(s_pop)
     state = st2;
     else if(s_read)
     state = st3;
     else
     state = st0;
     end

st1: begin // Reading data item to be pushed onto the stack
     intrnl_wr = 1'b1;
     intrnl_en = 1'b1;
     s_data_in = data_in;
     state = st4;
     end

st2: begin // Enabling stack memory to clear stack top
     intrnl_wr = 1'b0;
     intrnl_en = 1'b1;
     state = st5;
     end

st3: begin // Enabling stack memory to read data item
          // from stack top
     intrnl_wr = 1'b0;
     intrnl_en = 1'b1;
     state = st6;
     end

st4: begin
     intrnl_wr = 1'b0;
     intrnl_en = 1'b0;
     intrnl_add_bus =  intrnl_add_bus - 1'b1 ;
     state = st0;
     end

st5: begin
     intrnl_en = 1'b0;
     intrnl_add_bus =  intrnl_add_bus + 1'b1 ;
     state = st0;
     end

st6: begin
     intrnl_en = 1'b0;
     state = st0;
     end
```

```verilog
endcase
end
end

endmodule

/////////////////////////////////////////////////////////
// Module visted_node_store_128
//
// Description: Block RAM to store information on
// visited nodes
/////////////////////////////////////////////////////////

module visted_node_store_128 (clk,vis_node_en,vis_wr,
vis_node_add_bus,vis_node_in_data_bus,
vis_node_out_data_bus,vis_on) ;


// I/O declarations
input clk;
input  vis_node_en;// signal for enable
input vis_wr;
input [6:0] vis_node_add_bus;
input [7:0] vis_node_in_data_bus;
output [7:0] vis_node_out_data_bus;
output reg vis_on;
reg [7 :0] ram [127:0];
reg [6:0] read_a;

always @(posedge clk)
begin
if (vis_node_en)
begin
vis_on = 1'b1;
if (vis_wr)
ram[vis_node_add_bus] <= vis_node_in_data_bus;
read_a <= vis_node_add_bus;
end
else
vis_on = 1'b0;
end

assign vis_node_out_data_bus = ram[read_a];

endmodule


/////////////////////////////////////////////////////
```

```verilog
// Module uart
//
// Description:  Main module of UART to instantiate clock
// divider, sender  and control unit
//////////////////////////////////////////////////////////////

module uart(clk,rst,snd_out,com_rs,no_of_times_tr,cmd_gvn,
data,urt_state,send_state,data_out,stack_pop);

// I/O declarations
input clk,rst;
output snd_out; // Serial out signal
input [7:0] com_rs;
input [1:0] no_of_times_tr;
output cmd_gvn;
output [7:0] data;
output [3:0] urt_state;
output [3:0] send_state;
output [7:0] data_out;
input stack_pop;

wire dvdclk,req,ack,tx_rst,over,clk_dvd_rst;

// Instantiation of UART clock divider
uart_clk_divider divider0(clk,clk_dvd_rst,dvdclk);

// Instantiation of UART control unit
uart_cntrl cntrl0(clk,rst,tx_rst,clk_dvd_rst,req,ack,over,
data,rts,com_rs,no_of_times_tr,cmd_gvn,urt_state,
data_out,stack_pop);

// Instantiation of Universal Asynchronous Transmitter

uart_sender sender0(dvdclk,tx_rst,req,ack,over,data,
snd_out,send_state);

endmodule

//////////////////////////////////////////////////////////
// Module uart_clk_divider
//
// Description: Clock divider for UART to transmit
// data (commands) at the rate of few (about 10) Kbps
//////////////////////////////////////////////////////////


module uart_clk_divider (inClk,reset,outClk);

// I/O declarations
```

```verilog
input inClk;
input reset;
output outClk;
reg outClk;
reg [12:0] clockCount;

always @(posedge reset or posedge inClk)
begin
if (reset == 1'b1)
begin
clockCount = 13'b0000000000000;
outClk = 1'b0;
end
else
begin
clockCount = clockCount+1;
// assume a high Mhz input clock and count typically
// for a few  kHz output clock
if (clockCount == 2604)
begin
clockCount = 13'b0000000000000;
outClk = ~outClk;
end
end
end

endmodule

//////////////////////////////////////////////////////////////
// Module uart_cntrl
//
// Description: Control unit for UART (standard routine)
//////////////////////////////////////////////////////////////

module uart_cntrl (clk,rst,tx_rst,clk_dvd_rst,req,ack,over,
data,rts,com_rs,no_of_times_tr,cmd_gvn,urt_state,
data_out,stack_pop);

// I/O declarations
input clk,rst;
output req;
input ack;
input over;
output tx_rst;
output reg clk_dvd_rst;
output [7:0] data;
input rts;//request to send
input [7:0] com_rs;
input [1:0] no_of_times_tr;
```

```verilog
output reg cmd_gvn;
output [3:0] urt_state;
output [7:0] data_out;
input stack_pop;
reg tx_rst;
reg req;
reg [7:0] data;
reg [7:0] mem[0:2];
reg [3:0] state;
reg [1:0] counter;
integer i;
reg [3:0] int_cnt;
reg [1:0] rotation;
reg [31:0] delay_cnt;
reg turn;
reg frwd;

parameter st0=0,st1=1,st2=2,st3=3,st4=4,st5=5,
          st6=6,st7=7,st8=8;
assign urt_state = state;
assign data_out = data;

// Command generation

always @(posedge clk or posedge rst)
begin
if (rst == 1'b1)
begin
state = st0;
req = 1'b0;
mem[0] = 8'h00;
mem[1] = 8'h00;
mem[2] = 8'h00;
i=0;
tx_rst=1'b0;
clk_dvd_rst = 1'b0;
rotation = no_of_times_tr;
cmd_gvn = 1'b0;
end

else
begin
case(state)

st0: begin
    if(com_rs == 8'h6c || com_rs == 8'h72)
    begin
    frwd = 1'b1;
    turn = 1'b1;
```

```verilog
    int_cnt = 4'd5;
    mem[0] = com_rs;
    mem[1] = 8'h34;
    mem[2] = 8'h65;
    counter = 2'b0;
    end
    else
    begin
    frwd = 1'b0;
    turn = 1'b0;
    int_cnt = 4'd0;
    mem[0] = com_rs;
    mem[1] = 8'h31;
    mem[2] = 8'h65;
    counter = 2'b0;
    end
    clk_dvd_rst = 1'b1;
    tx_rst=1'b1;
    state = st1;
    end

st1: begin
    clk_dvd_rst = 1'b0;
    tx_rst=1'b0;
    req=1'b1;
    data=mem[i];
    state=st2;
    end

st2: begin
    if(!ack)
    state=st2;
    else
    begin
    req=1'b0;
    state=st3;
    end
    end

st3: begin
    if(!over)
    state=st3;
    else
    begin
    counter=counter+1'b1;
    i=i+1;
    tx_rst=1'b1;
    state=st4;
    end
```

```verilog
        end

   st4: begin
        if(counter < 2'b11)
        state=st1;
        else
        begin
        delay_cnt = 32'hfaf080f;
        state = st5;
        end
        end

   st5: begin
        if(delay_cnt == 0)
        state = st6;
        else
        begin
        delay_cnt = delay_cnt - 1;
        state = st5;
        end
        end

   st6: begin
        if(int_cnt > 4'd2)
        begin
        int_cnt = int_cnt - 4'b0001;
        counter = 2'd0;
        i=0;
        state = st1;
        end
        else if(int_cnt == 4'd2 && frwd == 1'b1)
        begin
        int_cnt = int_cnt - 4'b0001;
        mem[0] = 8'h62;
        mem[1] = 8'h31;
        counter = 2'd0;
        i=0;
        state = st1;
        end
        else if(turn == 1'b1)
        state = st8;
        else
        state = st7;
        end

   st7: begin
        cmd_gvn = 1'b1;
        state =st7;
        end
```

```
st8: begin
     turn = 1'b0;
     frwd = 1'b0;
     int_cnt = 4'd5;
     if(com_rs == 8'h6c)
     mem[0] = 8'h72;
     else
     mem[0] = 8'h6c;
     mem[1] = 8'h34;
     mem[2] = 8'h65;
     counter = 2'b0;
     i=0;
     state =st1;
     end
endcase
end
end
endmodule




/////////////////////////////////////////
// Module ultra_sonic_sensor
//
// Description: The Pulse Width to Distance Converter
//////////////////////////////////////////////

module ultra_sonic_sensor(clk,ultra_rqst,sen_out,
dist_val,sen_over);

// I/O signals
input clk;
input ultra_rqst;
input sen_out;
output [15:0]  dist_val;
output sen_over;
wire cnt_en;
wire out_clk;
wire [15:0] count_out;

// Instantiation of counter for measuring distance
// to the nearest obstacles
ultra_sen_counter cnt0(count_out,cnt_en,out_clk,ultra_rqst);

// Instantiation of control unit for the Pulse Width to
//  Distance converter
ultra_sen_cntrl s0(cnt_en,clk,sen_out,ultra_rqst,count_out,
```

```verilog
dist_val,sen_over);

// Instantiation of clock divider for Pulse Width to
// Distance Converter
ultra_sens_clkdivider ckdv0(clk,out_clk,ultra_rqst);

endmodule

//////////////////////////////////////////////////
// Module ultra_sen_cntrl
//
// Description: The control unit of pulse width
// to distance converter
//////////////////////////////////////////////////////

module ultra_sen_cntrl(count,clk,pin,ultra_rqst,count_out,
dist_val,sen_over);

output count;    //enable signal to the counter
input clk, pin, ultra_rqst;// pin is the output of the
                           // ultrasonic sensor
output [15:0] dist_val;
input [15:0] count_out;
output reg sen_over;
reg count;
parameter   s0=0,s1=1,s2=2,s3=3,s4=4;
reg [2:0] state;
reg [6:0]int_sig;
reg [15:0] str_dist;


assign dist_val = {5'b0,count_out[15:5]};
always @(posedge clk or negedge ultra_rqst)
begin

if(!ultra_rqst)
begin
count = 0;
state = s0;
int_sig = 7'b0;
sen_over = 1'b0;
str_dist = 16'b0;
end
else
begin

case(state)

//wait for logic zero of pin
```

```verilog
s0: begin
    if(~pin)
    begin
    count = 0;
    state = s1;
    end
    else
    begin
    count = 0;
    state = s0;
    end
    sen_over = 1'b0;
    end

//wait for logic one of pin
s1: begin
    if(pin)
    begin
    count = 0;
    state = s2;
    end
    else
    begin
    count = 0;
    state = s1;
    end
    end

// Give enable signal to the counter during the logic
// high of pin
s2: begin
    if(pin)
    begin
    count = 1;
    state = s2;
    end
    else
    begin
    count = 0;
    int_sig = int_sig + 1'b1;
    state = s3;
    end

// dummy state to ignore subsequent pulses coming from sensor
s3: begin
    if(int_sig < 7'd32)
    begin
    count = 0;
    state =s0;
```

```verilog
        end
        else
        begin
        count = 0;
        sen_over = 1'b1;
        state = s3;
        end
        end

    s4: begin
        state = s4;
        sen_over = 1'b0;
        end

    endcase
    end
    end

    endmodule


    //////////////////////////////////////////////////
    // Module ultra_sen_counter
    //
    // Description:  Counter for measuring the distance
    ///////////////////////////////////////////////////////

    module ultra_sen_counter(count_out,count,outclk,ultra_rqst);

    output [15:0]count_out; //contains distance to the
                            // nearest obstacle
    input outclk,  //output of the clock divider(.176 MHz)
        ultra_rqst,
        count;   //enable signal for the counter
    wire int_sig;
    reg [15:0]count_out;


    always @ (posedge outclk or negedge ultra_rqst)
    if(!ultra_rqst)
    count_out = 16'b0;
    else
    if(count)
    count_out = count_out + 1'b1;

    endmodule
```

```verilog
//////////////////////////////////////////////////////////
// Module ultra_sens_clkdivider
//
// Description: Clock divider for pulse width to
//                 distance converter
//////////////////////////////////////////////////////////

module ultra_sens_clkdivider(clk,outclk,ultra_rqst);

// I/O declarations
  input clk,                        //input clock (50 MHz)
         ultra_rqst;
  output outclk;                    //output clock (.176 MHz)
  reg outclk;
  reg [7:0]cnt;
  wire int_sig;

always @ (posedge clk or negedge ultra_rqst)
begin
if(!ultra_rqst)
begin
cnt=8'h0;
outclk=0;
end
else
if(cnt == 0)
begin
outclk = ~ outclk; //toggle counter after cnt becomes zero
cnt = 8'h8e;       // load counter with 142
end       // (50MHz/284 = .176 MHz)
else
cnt = cnt - 1;
end

endmodule
```

B

Suggestions for Mini-Projects

This appendix presents a list of mini-projects for senior undergraduate and graduate students in various disciplines of engineering. An FPGA board of the kind described in this book or one with similar capabilities (in terms of support for display of output and interfaces) is required for these projects.

Project 1: Development of Low-Cost Ultrasonic Sensors

This mini-project is based on ideas presented in Chapter 3. The project consists of (i) Using readily available TR40 transducers and ICs to implement the ultrasonic sensor initially on a breadboard and (ii) Testing the sensor using an oscilloscope (preferably a digital storage oscilloscope) on objects such as cardboard boxes placed at varying distances from the transducers.

It would be of interest to observe the behaviour of the sensors for other types of objects such as plastic cans, glass bottles etc. of varying dimensions. A printed circuit board-based solution can be developed as the last part of this project.

Project 2: Processing Ultrasonic Sensor Data Using an FPGA

The objective of this mini-project is to build up on the first by designing a digital system to process sensor input. This would involve (i) Development of programs (in a hardware description language) for converting pulse width into an n-bit number (n could be chosen as 8 or 16) and (ii) Verification of the output on seven-segment displays (provided on the FPGA board). Basic ideas for the design are available in Chapter 4.

Project 3: A 4-bit Stack-based Microprocessor on an FPGA

This mini-project is intended to build up on the stack memory programs (in Verilog) given in this book. The requirement is to design and implement a 4-bit stack based microprocessor that can be used for issuing a set of commands to a robotic vehicle. The processor has to perform addition, subtraction and multiplication operations based on PUSH and POP operations The implementation can be tuned to fit on a low-end FPGA to make the solution cost-effective and space-efficient. Effective utilization of the FPGA resources (in terms of LUTs, block RAMs etc.) would be of interest in this project.

Project 4: Universal Asynchronous Transmitter and Receiver Design

The goal of this project is to make use of the RS-232 serial interface available on an FPGA board for communication with a PC. The design should have a UART (Universal Asynchronous Receiver and Transmitter), an 8-bit input bus and an 8-bit output bus. The details of UAT (Universal Asynchronous Transmitter) given in this book can be used for designing the UART. The 8-bit input bus is used for receiving a character from the PC and the 8-bit output data bus is used to display character transmitted by the FPGA on the seven segment displays on the FPGA board.

Project 5: Power Measurement and Selective Shutdown

The first objective in this mini-project is to get exposure to power measurement tools provided by Xilinx and other vendors. In particular, *X Power* and Xilinx SpartanIIE power estimate worksheet can be studied for measuring power consumed by modules starting from simple logic elements such as adders and counters.

The second objective is to understand how selective shutdown works (based on ideas presented in Chapter 4) and when it is advantageous to incorporate additional circuitry for selective shutdown. Starting from simple elements like adders, one can proceed to more complex elements such as multipliers and trigonometric function evaluators to see the power consumption with and without selective shutdown. The area overhead is also worth examining. Code presented in Appendix 1 illustrating use of *enable signals* would be valuable for this study.

Project 6: FPGA-based Stepper Motor Control

The objective of this project is control of the rotation of a stepper motor using an FPGA. The design should have a 4-bit data bus which is connected to the 4 inputs of a stepper motor driver card. The stepper motor driver card has four output lines corresponding to the four phases which are connected appropriately to the stepper motor.

Project 7: Barcode Reader Interface to an FPGA

A simple mechanism for locating a robot in an indoor environment is using a barcode reader. Labels containing barcodes can be pasted on the floor and the codes may represent coordinates of different points in the environment with respect to the origin located in some corner.

In this project, we assume a barcode reader is available (A number of low cost barcode readers are available in the market and many come with support to keep the "beam" on continuously when the reader is mounted on a robot). Interfacing a barcode reader to an FPGA and processing the input from a barcode reader are the objectives of this project. It would be valuable to download software for generating labels containing barcodes. Freeware is available on the Internet for this purpose.

As an extension to this exercise, the orientation of the labels containing the barcodes can be changed and the effectiveness of scanning in the codes can be examined at different distances from the floor (on which the codes are pasted).

Project 8: Parallel Processing of Ultrasonic Sensor Data by an FPGA

The research presented in this book has brought out the value of *parallel processing* of sensor data by an FPGA. The objective of this mini-project is to test how a set of two or more ultrasonic sensors interfaced to an FPGA perform when objects in the environment are dynamic. Two or more humans can move simultaneously in the vicinity of the ultrasonic sensors (analogous to the scenarios depicted earlier) and the seven segment display on the FPGA board can be used to show the count of the number of moving objects.

Closing remarks

The various miniprojects are intended to give a feel for various aspects of FPGA-based robotic system design and experimentation.

References

1. J.T. Schwartz and C.K. Yap. *Algorithmic and Geometric Aspects of Robotics.* Lawrence Erlbaum Associates, 1987.
2. T. Lozano-Perez and M.A. Wesley. An algorithm for planning collision-free paths among polyhedral obstacles. *Communications of the ACM*, 22(10):560–570, 1979.
3. F. Preparata and M.I. Shamos. *Computational Geometry: An Introduction.* Springer-Verlag, 1985.
4. H. Choset and J. Burdick. Sensor-based exploration: the hierarchical generalized Voronoi graph. *International Journal of Robotics Research*, pages 96–125, 2000.
5. V. J. Lumelsky and A. A. Stepanov. Path-Planning Strategies for a Point Mobile Automaton. *Algorithmica*, 2:403–430, 1987.
6. E. Magid and E. Rivlin. CAUTIOUSBUG: a competitive algorithm for sensory-based robot navigation. In *Proceedings of 2004 IEEE/RSJ International Conference on Intelligent Robots and Systems*, pages 2757–2762, 2004.
7. J. Evans, B. Krishnamurthy, B. Barrows, T. Skewis, and V. Lumelsky. Handling real-world motion planning: a hospital transport robot. *IEEE Control Systems*, pages 15–19, 1992.
8. Y. Mei, Y-H. Lu, Y.C. Hu, and C.S.G. Lee. A case study of mobile robot's energy consumption and conservation techniques. In *Proceedings of Int. Conf. on Advanced Robotics*, pages 492–497, 2005.
9. A. DeHon. The density advantage of reconfigurable computing. *IEEE Computer*, 33:41–49, 2000.
10. N.J. Nilsson. A mobile automation: an application of artificial intelligence techniques. In *Proceedings of Int. Joint Conf. on Artificial Intelligence*, 1969.
11. H. Everett. A microprocessor controlled autonomous sentry robot. Master's thesis, Naval Postgraduate School, 1982.
12. A. Meyrowitz, R. Blidberg, and R.C. Michelson. Autonomous vehicles. *Proceedings of the IEEE*, 84(8):1147–1164, 1996.
13. J. Xiao, A. Calle, J. Ye, and Z. Zhu. A mobile robot platform with DSP-based controller and omnidirectional vision system. In *Proceedings of IEEE International Conference on Robotics and Biomimetics*, pages 844–848, 2004.
14. V. Gabbani, S. Rocchi, and V. Vignoli. Multielement ultrasonic system for robotic navigation. In *Proceedings of IEEE International Conference on Robotics and Automation*, pages 3009–3014, 1995.

15. A. Elfes. Sonar-based real-world mapping and navigation. *IEEE Journal of Robotics and Automation*, RA-3(3):249–265, 1987.

16. L. Kleeman. Ultrasonic autonomous robot localisation system. In *Proceedings of IEEE International Workshop on Intelligent Robots and Systems*, pages 212–219, 1989.

17. C.C. Tong, J.F. Figueroa, and E. Barbieri. A method for short or long range time-of-flight measurements using phase-detection with an analog circuit. *IEEE Transactions on Instrumentation and Measurement*, 50:1324–1328, 2001.

18. M. Beckerman and E.M. Oblow. Treatment of systematic errors in the processing of wide-angle sonar sensor data for robotic navigation. *IEEE Transactions on Robotics and Automation*, 6:137–145, 1990.

19. H. Peremans, K. Audenaert, and J.M. Van Campenhout. A high resolution sensor based on tri-aural perception. *IEEE Transactions on Robotics and Automation*, 9:36–48, 1993.

20. L. Korba, S. Elgazzar, and T. Welch. Active infrared sensors for mobile robots. *IEEE Transactions on Instrumentation and Measurement*, 43:283–287, 1994.

21. D. Jung and A. Zelinsky. Whisker based mobile robot navigation. In *Proceedings of IEEE International Conference on Intelligent Robots and Systems*, pages 497–504, 1996.

22. G.N. DeSouza and A. Kak. Vision for mobile robot navigation: A survey. *IEEE Transactions on Pattern Analysis and Machine Intelligence*, 24:237–267, 2002.

23. J.C. Latombe. *Robot Motion Planning*. Kluwer Academic Publishers, 1991.

24. B. Kuipers and Y.T. Byun. A robot exploration and mapping strategy based on a semantic hierarchy of spatial representations. *Robotics and Autonomous Systems*, 8:47–63, 1991.

25. G. Dudek, M. Jenkin, E. Milios, and D. Wilkes. Robotic exploration as graph construction. *IEEE Transactions on Robotics and Automation*, 7:859–865, 1991.

26. N.S.V. Rao, N. Stolzfus, and S.S. Iyengar. A retraction method for learned navigation in unknown terrains for a circular robot. *IEEE Transactions on Robotics and Automation*, 7:699–707, 1991.

27. H. Choset, S. Walker, K. Eiamsa-Ard, and J. Burdick. Sensor-based exploration: Incremental construction of the hierarchical generalized Voronoi graph. *The International Journal of Robotics Research*, 19(2):126–148, 2000.

28. H. Choset, K.M. Lynch, S. Hutchinson, G. Kantor, W. Burgard, L.E. Kavraki, and S. Thrun. *Principles of robot motion: Theory, algorithms, and implementations*. MIT Press, 2005.

29. R. Mahkovic and T. Slivnik. Constructing the generalized local Voronoi diagram from laser range scanner data. *IEEE Transactions on Systems, Man and Cybernetics - Part A: Systems and Humans*, 30(6):710–719, 2000.

30. I. Kamon, E. Rivlin, and E. Rimon. A new range-sensor based globally convergent algorithm for mobile robots. In *Proc. IEEE International Conf. on Robotics and Automation*, pages 429–435, 1996.

31. C. Taylor and D. Kriegman. Exploration strategies for mobile robots. In *Proceedings of the IEEE International Conference on Robotics and Automation*, pages 248–253, 1993.

32. H.H. Gonzalez-Banos and J.C. Latombe. Navigation strategies for exploring indoor environments. *International Journal of Robotics Research*, 21:829–848, 2002.

33. R. Biswas, B. Limketkai, S. Sanner, and S. Thrun. Towards object mapping in non-stationary environments with mobile robots. In *Proceedings of Int. Conf. on Intelligent Robots and Systems*, pages 1014–1019, 2002.

34. H. Zha, K. Tanaka, and T. Hasegawa. Detecting changes in a dynamic environment for updating its maps by using a mobile robot. In *Proceedings of Int. Symp. on Intelligent Robots and Systems*, pages 1729–1734, 1997.

35. T. Duckett and U. Nehmzow. Mobile robot self-localisation and measurement of performance in middle-scale environments. *Robotics and Autonomous Systems*, 24:57–69, 1998.

36. B. Tovar, L.M. Gomez, R.M. Cid, M.A. Miranda, R. Monroy, and S. Hutchinson. Planning exploration strategies for simultaneous localization and mapping. *Robotics and Autonomous Systems*, 54:314–331, 2006.

37. X. Deng, E. Milios, and A. Mirzaian. Landmark selection for path execution. In *Proceedings of 1993 IEEE/RSJ International Conference on Intelligent Robots and Systems*, pages 1339–1346, 1993.

38. J. Horn and G. Schmidt. Continuous localization for long-range indoor navigation of mobile robots. In *Proceedings of 1995 IEEE International Conference on Robotics and Automation*, pages 387–394, 1995.

39. B. Yamauchi and R. Beer. Spatial learning for navigation in dynamic environments. *IEEE Transactions on Systems, Man, and Cybernetics*, 26:496–505, 1996.

40. R. Greiner and R. Isukapalli. Learning to select useful landmarks. *IEEE Transactions on Systems, Man and Cybernetics - Part B: Cybernetics*, 26:437–448, 1996.

41. I. Moon, J. Miura, and Y. Shirai. On-line selection of stable visual landmarks under uncertainty. In *Proceedings of 1999 IEEE/RSJ International Conference on Intelligent Robots and Systems*, pages 781–786, 1999.

42. J.A. Castellanos, M. Devy, and J.D. Tardós. Towards a topological representation of indoor environments: A landmark-based approach. In *Proceedings of 1999 IEEE/RSJ International Conference on Intelligent Robots and Systems*, pages 23–28, 1999.

43. S. Thrun. Bayesian landmark learning for mobile robot localization. *Machine Learning*, 33:41–76, 1998.

44. C.F. Olson. Landmark selection for terrain matching. In *Proceedings of 2000 IEEE International Conference on Robotics and Automation*, pages 1447–1452, 2000.

45. R.H. Luke, J.M. Keller, M. Skubic, and S. Senger. Acquiring and maintaining abstract landmark chunks for cognitive robot navigation. In *Proceedings of 2005 Int. Conf. on Intelligent Robots and Systems*, pages 2566–2571, 2000.

46. J. Borenstein and Y. Koren. The vector-field histogram - fast obstacle avoidance for mobile robots. *IEEE Transactions on Robotics and Automation*, 7:278–288, 1991.

47. R. Biswas, B. Limetkai, S. Sanner, and S. Thrun. Towards object mapping in non-stationary environments with mobile robots. In *Proceedings of the IEEE/RSJ Conference on Intelligent Robots and Systems*, pages 1014–1019, 2002.

48. C. Stachniss and W. Burgard. Mapping and exploration with mobile robots using coverage maps. In *Proceedings of the IEEE/RSJ Conference on Intelligent Robots and Systems*, pages 467–472, 2003.

49. E. Prassler, J. Scholz, and P. Fiorini. Navigating a robotic wheelchair in a railway station during rush hour. *The International Journal of Robotics Research*, 18(7):711–727, July 1999.

50. N. Ranganathan, B. Parthasarathy, and K. Hughes. A parallel algorithm and architecture for robot path planning. In *Proceedings of 1994 IEEE Eighth Int. Parallel Processing Symposium*, pages 275–279, 1994.

51. P. Tzionas, A. Thanailakis, and P. Tsalides. Collision-free path planning for a diamond shaped robot using two-dimensional cellular automata. *IEEE Transactions on Robotics and Automation*, 13:237–250, 1997.

52. N. Sudha and K. Sridharan. A high speed VLSI design and ASIC implementation for constructing Euclidean distance based discrete Voronoi diagram. *IEEE Transactions on Robotics and Automation*, 20(2):352–358, 2004.

53. K. Hoff III, T. Culver, J. Keyser, M.C. Lin, and D. Manocha. Fast computation of generalized Voronoi diagrams using graphics hardware. In *Proceedings of ACM SIGGRAPH*, pages 1844–1845, 1999.

54. A. Sud, E. Andersen, S. Curtis, M. Lin, and D. Manocha. Real-time path planning for virtual agents in dynamic environments. In *Proceedings of IEEE Virtual Reality Conference*, pages 91–98, 2007.

55. J. Janet, R.C. Luo, and M.G. Kay. Autonomous mobile robot motion planning and geometric beacon collection using traversability vectors. *IEEE Transactions on Robotics and Automation*, 13(1):132–140, 1997.

56. S.K. Lam and T. Srikanthan. High speed environment representation scheme for dynamic path planning. *Journal of Intelligent and Robotic Systems*, 32:307–319, 2001.

57. K. Sridharan and T.K. Priya. The design of a hardware accelerator for real-time complete visibility graph construction and efficient FPGA implementation. *IEEE Transactions on Industrial Electronics*, 52:1185–1187, 2005.

58. Lam Siew Kei, K. Sridharan, and T. Srikanthan. VLSI-efficient schemes for high-speed construction of tangent graph. *Robotics and Autonomous Systems*, 51:248–260, 2005.

59. D.S. Katz and R.R. Some. NASA advances robotic space exploration. *IEEE Computer*, 36, 2003.

60. C.R. Lee and Z. Salcic. High-performance FPGA-based implementation of Kalman filter. *Microprocessors and Microsystems*, 21:257–265, 1997.

61. G.J. Gent, S.R. Smith, and R.L. Haviland. An FPGA-based custom coprocessor for automatic image segmentation applications. In *Proc. of IEEE Workshop on FPGAs for Custom Computing Machines*, pages 172–179, 1994.

62. M. Alderighi, F. Casini, S. D'Angelo, D. Salvi, and G.R. Sechi. A fault-tolerant FPGA-based multi-stage interconnection network for space applications. In *Proceedings of IEEE Int. Workshop on Electronic Design, Test and Applications*, pages 302–306, 2002.

63. S.J. Visser, A.S. Dawood, and J.A. Williams. FPGA based real-time adaptive filtering for space applications. In *Proceedings of IEEE Int. Conf. on Field-Programmable Technology*, pages 322–326, 2002.

64. M.S. Masmoudi, I. Song, F. Karray, M. Masmoudi, and N. Derbel. FPGA implementation of fuzzy wall-following control. In *Proceedings of the 16th IEEE International Conference on Microelectronics*, pages 133–139, 2004.

65. Q. Zhou, K. Yuan, H. Wang, and H. Hu. FPGA-based colour image classification for mobile robot navigation. In *Proceedings of the IEEE International Conference on Industrial Technology*, pages 921–925, 2005.

66. Y. Meng. A dynamic self reconfigurable robot system. In *Proceedings of IEEE International Conference on Advanced Intelligent Mechatronics*, pages 1541–1546, 2005.

67. C. Paiz, T. Chinapirom, U. Witkowski, and M. Porrmann. Dynamically reconfigurable hardware for autonomous mini robots. In *Proceedings of IEEE 32nd Annual Conference on Industrial Electronics*, pages 3981–3986, 2006.

68. V. Tadigotla, L. Sliger, and S. Commuri. FPGA implementation of dynamic runtime behavior of reconfiguration in robots. In *Proceedings of IEEE International Symposium on Intelligent Control*, pages 1220–1225, 2006.

69. AN470 - Application Note. In *www.st.com*, 2003.

70. J.W. Jang, S.B. Choi, and V.K. Prasanna. Energy and time-efficient matrix multiplication on FPGAs. *IEEE Transactions on VLSI Systems*, 13(11):1305–1319, 2005.

71. T.H. Cormen, C.E. Leiserson, and R.L. Rivest. *Introduction to Algorithms*. MIT Press, 1990.

72. T. Ogura, S. Yamada, and T. Nikaido. A 4-kbit associative memory LSI. *IEEE Journal of Solid State Circuits*, SC-20(6):1277–1282, Dec. 1985.

73. T. Ogura, S. Yamada, and J. Yamada. A 20-kbit CMOS associative memory LSI for artificial intelligence machines. In *Proceedings of IEEE Int. Conf. Computer Design: VLSI in Computers*, pages 574–577, 1986.

74. J.P. Hayes. *Computer Architecture and Organization*. McGraw-Hill, 1998.

75. L. Chisvin and R.J. Duckworth. Content-addressable and associative memory: Alternatives to the ubiquitous RAM. *IEEE Computer*, pages 51–64, 1989.

76. V. Tiwari, S. Malik, and P. Ashar. Guarded evaluation: Pushing power management to logic synthesis/design. *IEEE Transactions on Computer-Aided Design of Integrated Circuits and Systems*, 17(10):1051–1060, Oct. 1998.

77. Jan M. Rabaey, A. Chandrakasan, and B. Nikolic. *Digital Integrated Circuits*. Prentice-Hall, 2003.

78. XAPP173 - Using BlockSelectRAM+Memory in Spartan-II FPGAs. In *www.xilinx.com*, 2000.

79. S. Choi, R. Scrofano, V.K. Prasanna, and J.-W. Jang. Energy-efficient signal processing using FPGAs. In *Proceedings of Eleventh ACM International Symposium on Field Programmable Gate Arrays*, pages 225–234, 2003.

80. P. Rajesh Kumar, K. Sridharan, and S. Srinivasan. A parallel algorithm, architecture and FPGA realization for landmark determination and map construction in a planar unknown environment. *Parallel Computing*, pages 205–221, 2006.

81. J.M. O'Kane and S.M. LaValle. Localization with limited sensing. *IEEE Transactions on Robotics*, pages 704–716, 2007.

Index

MIX
Papier aus verantwortungsvollen Quellen
Paper from responsible sources
FSC® C105338

If you have any concerns about our products,
you can contact us on
ProductSafety@springernature.com

In case Publisher is established outside the EU,
the EU authorized representative is:
Springer Nature Customer Service Center GmbH
Europaplatz 3, 69115 Heidelberg, Germany

Printed by Libri Plureos GmbH
in Hamburg, Germany